Digital Fabricators

Curator
Michael Stacey Building Centre Trust and London Metropolitan University
with **Philip Beesley** and **Vincent Hui,** University of Waterloo

cambridge **DESIGN AT RIVERSIDE**

Cambridge Galleries
November 11, 2004 - January 30 2005

in association with the
2004 AIA/ACADIA Fabrication Conference

Hosted by
University of Waterloo
School of Architecture in Cambridge
with
University of Toronto AL+D

1

Design and Production:
Philip Beesley, Farid Noufaily,
Marianne Magus, Jonathan Tyrrell, Shane Williamson

Produced in cooperation with Cambridge Galleries Design at Riverside
Mary Misner, Director and John McMinn, Coordinator

In association with the
AIA/ ACADIA Fabrication conference,
University of Waterloo Cambridge and University of Toronto AL+D

Printing by Coach House Press, Toronto, Ontario

Library and Archives Canada 1-897001-03-7
Digital Fabrication
Michael Stacey, curator
with Philip Beesley and Vincent Hui

Digital Fabricators
Curator
Michael Stacey Building Centre Trust and London Metropolitan University
with Philip Beesley and Vincent Hui, University of Waterloo

ISBN 1-897001-03-7
1. Architectural Design
2. Architectural Practice
3. Building Construction

I. Michael Stacey
II. Philip Beesley
III. Vincent Hui

Mary Misner
Director, Cambridge Galleries

Cambridge Galleries are pleased to present the exhibition, *Digital Fabricators* in conjunction with the AIA/ACADIA Fabrication Conference, hosted by the University of Waterloo School of Architecture and the University of Toronto School of Architecture, November 8 to 13, 2004.

Curated by Michael Stacey and the *Digital Fabricators* Research Group at London Metropolitan University with the Building Centre Trust, the Digital Fabricators exhibition is prominent in our exhibition schedule not only for the vision put forward by its curators, but because it is one of the inaugural exhibitions in our new gallery space, Design at Riverside.

This gallery is dedicated to the exhibition of architecture and design and we look forward to working with the University of Waterloo School of Architecture in accomplishing this mandate.

Working with Philip Beesley, Vincent Hui and the Conference Committee on the preparations for the *Digital Fabricators* exhibition has helped us establish this partnership. We are most grateful for their cooperation and goodwill. Our special acknowledgement goes to Michael Stacey for his curatorial wisdom and insightful essay.

We welcome this extraordinary exhibition and the opportunity to share it with a wider audience.

Digital Fabricators

Curated by Michael Stacey and the Digital Fabricators Research Group

 VS.

This exhibition explores the relationship between architecture, manufacturing techniques and digital technology.

The landscape of every architect's office has changed over the past 20 years. Gone is the gentle squeak of Rotring pen on Mylar or tracing paper to be replaced by the hum of computers and the intense clicking of mice. This change has been embraced by architects and engineers. But two-dimensional drafting still dominates the construction industry and is used primarily for its flexibility and a hoped-for efficiency. The revolutionary potential of three-dimensional modelling is used fitfully and only by a few. This article, like the exhibition, explores the use of digital design to inform the built environment. The emphasis is on experiential and tactile architecture, not the theoretical.

The potency of the sketch and the three-dimensional models held in an architect's imagination are beyond doubt. But the communicative potential of digital design, in all stages of the design process from concept to direct communication with the fabricators, is still in its infancy in the construction industry.

The exhibition includes the timeline of the development of digital design and fabrication technology. For example, the introduction of AutoCAD in 1982 had a much swifter impact than the introduction of stereolithography in 1988. The projects illustrate the expanding diversity of digital fabricating techniques, from laser cutting to five-axis routing, and stereolithography to three-dimensional physical printing. The exhibition includes a taxonomy of current digital fabrication technology. The following classifications of digital fabrication have been used: two-dimensional, subtractive, additive and formative. Within each category there is a delta of possibilities, and many of the fundamental issues of tectonics in architecture, the joining of materials and components, remain unchanged by the use of digital fabrication. This is leading to a re-engagement with the means of production by the profession and a rediscovery of craft in architecture. Additive processes including laser sintering, stereolithograph and three-dimensional printing produced components that appear to be the products of subtractive sculpture but are all formed topographically layer-by-layer. Thus the conventional sculptural distinction between constructional and subtraction has been totally subverted.

Some are concerned that rapid prototyping in particular lacks any sense of materiality; however, Ulrika Karlsson of Servo notes that the layered topography of the stereolithography rapid prototypes of lattice archipelogics has it *own and unique materiality, which is a direct result of the setting of the resin by exposure to laser light, layer by layer.* Laser cutting uses the nature of the chosen material directly – Philip Beesley and Diane Willow's Orpheus Filter is an accretive installation formed from laser-cut acrylic and Mylar film. Thus they are using the very film one used to draw on to create working drawings.

Parametric design

The combination of parametric design and single project models offers the architect a potent real-time tool to generate options and iterate the design to access the potential within a conceptual approach. Parametrics define the parameters of a particular design and not its shape. This is a powerful new tool in form-finding for architecture. A parametric definition of a circle is $r2 = x2 + y2$, and the parametric definition of the arch of Waterloo Station as defined

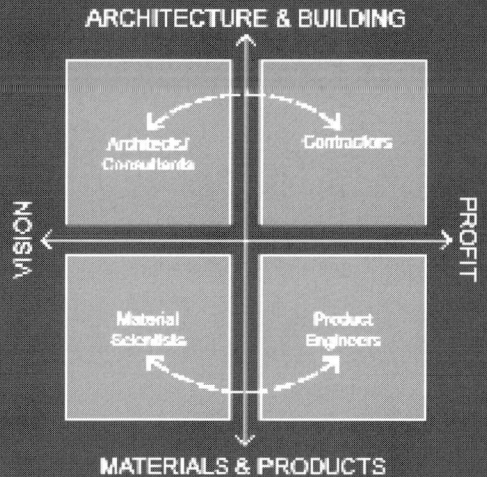

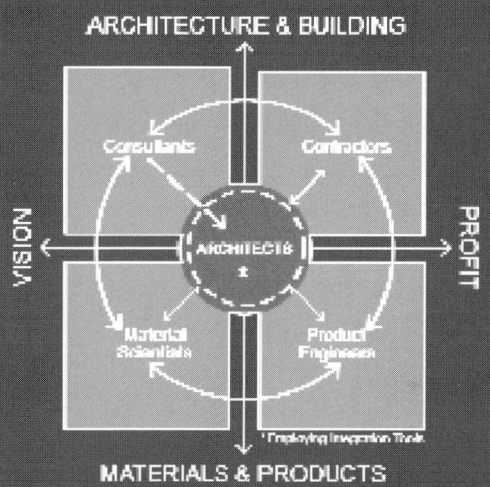

by Robert Aish of Bentley Systems as $hx = ((29152 + (B+C)2)0.5)$.

This is not to suggest that practices should all hire mathematicians, which Foster has done, nor that you should enrol for a maths degree. Thankfully, major software companies are developing visual interfaces or 'self-programming' for parametric design. The parametric capability within Bentley's Microstation suite is called 'generative components'.

The Swiss Re office building is a pioneering exemplar of parametric design. It takes the market preconception of an office layout and, through parametrics, transforms it into an optimal design where aesthetics emerge from performance criteria applied to design. Francis Aish, an aerospace engineer by training and now an associate at Foster and Partners, describes this parametric process as 'two and half D' as the parametric qualities of the seven tangential arcs that form the profile were applied to the sectional geometry and related to the circular plan.

One does not need to be the architect for a regional arts centre or major office building to use parametric design. Urban Future Organization's design for a house conversion in London called Nested House V1.03 ably demonstrates this on a budget of under £60,000. It is also vital to remember that the development of digital design is about the investment in people – Gehry transformed his office by hiring Jim Glymph and Rick Smith. The realisation of Foster projects such as the GLA and Swiss Re is only possible because of the investment in key software skills, in people, by fabricators such as Wagner Biro and Schmidlin.

Foster versus Gehry

In the use of software there is a strong contrast between the approach of Gehry and Foster. Gehry has adopted an approach inspired by Boeing and uses CATIA (computer aided three-dimensional interactive applications). Furthermore, it requires the complete supply chain to adopt this single software to ensure continuity and compatibility. Foster's approach on Swiss Re, however, was to seek a robust software that everyone had access to – Excel. The geometry of the project was communicated as an Excel spreadsheet

and a method statement on how to 'generate' the geometry. The specialist subcontractors' resultant geometry was then inspected by Foster's design team and any divergence discussed and eliminated. Swiss Re also very excellently demonstrates the interaction of physical models, made by the architects, and their digital models, a flip-flop from the physical to digital and back again, until all the consequences of the geometry are fully understood. It is salient to note that as soon as true depth is applied to a specific geometry, for example the straight line or ruled geometry of the hyperbolic paraboloid roofs of Richard Rogers Partnership's Antwerp Law Courts, a curvilinear geometry is encountered. This was resolved by the close collaboration of Avtar Lotay, the project architect, and the specialist timber fabricator Merk. On Swiss Re, Foster's design team resolved the geometry to clad it exclusively with flat quadrilateral panels with the exception of the crowning double-curved rooflight at the apex of the tower.

When reviewing the progression from digital representation to digital fabrication, it is easy to overemphasise the importance of technology transfer from the aerospace and automotive industries. This is not to deny the importance of software such as CATIA – however, progress within the construction industry, which includes CIMsteel, should not be overlooked. It is also pertinent to note that CIMsteel was a European initiative, which has now been taken up in North America.

No other industry is willing to undertake the risk of putting its prototypes on street corners and then standing back for, say, 20 to 30 years to see what happens. The building industry at its best is capable of putting together flexible project teams with disparate skills in the pursuit of common project goals. The use of digital fabrication technology need not be part of a more corporate future – the dialogue with industry should be a two-way process.

The possibilities generated by the direct manufacture of digital designs are both an opportunity and a challenge to the architectural profession. The technology is accessible and cost-effective. The inventive skills and editorial judgement of architects are needed more than ever.

The *Digital Fabricators* exhibition was curated by Michael Stacey and the Digital Fabricators Research Group at London Metropolitan University with the Building Centre Trust. Digital Fabricators North American stage of the exhibition was co-curated by Philip Beesley and Vincent Hui in cooperation with the Cambridge Art Gallery and the AIA/ACADIA Fabrication Conference.

Venues were:
Interbuild Birmingham 25-29 April, Building Centre London 10 May-26 June, and University of Waterloo, Cambridge Galleries and University of Toronto AL+D, Canada, from 11 November 2004-29 January 2005.

"Twenty years ago it was common for engineers to spend long tedious hours working out the way in which a two-storey building frame worked, longhand, on paper. Sometimes, ideas were incidental to the process. Thankfully, those days are past. The arrival of interactive design software has revolutionised the way we design things… It means that engineering has become more of an art, architecture more of a science, and all design more intuitive' (Professor Chris Wise)

The Sagrada Familia

Mark Burry

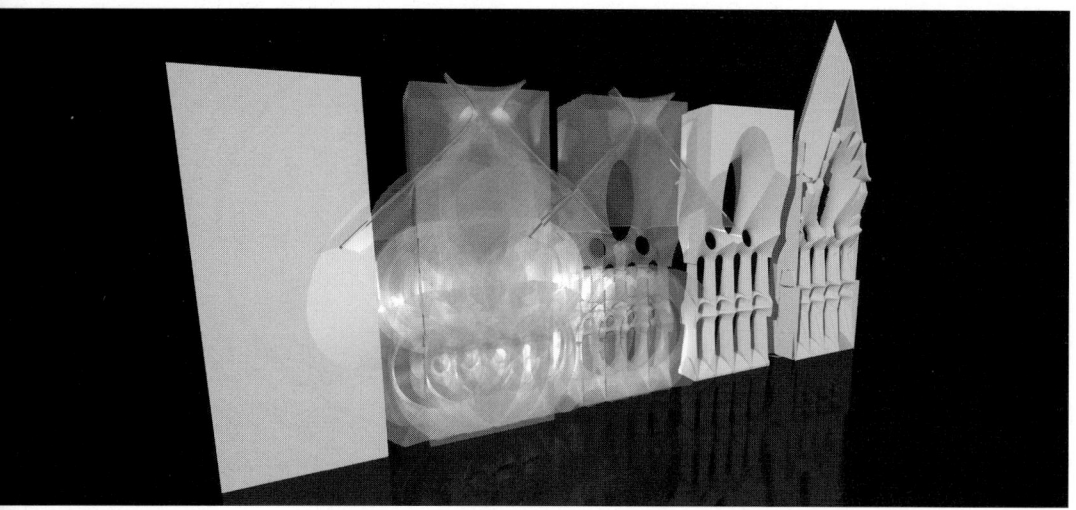

Parametric modelling allowed Australian and Spanish teams to collaborate on the fast-track design and construction of the west transept rose window at Antoni Gaudì's unfinished Sagrada Familia.

The recently redesigned west transept rose window ('passion facade') of Antoni Gaudì's unfinished major work and Barcelona icon, the Sagrada Familia church, was designed and completed in a little over 12 months and is a notable example of 'lean construction'. The processes involved include traditional stone masonry, actual employment of the traits discussed in Robin Evans' The Projective Cast, and semi-automated construction methods.

Parametrics

Working in the technical office on site, Mark Burry constructed a flexible model of the whole assembly by using parametric shipbuilding software, called CADDS5. The model, which contained a total of 3,800 events and took six weeks to build, uses associative geometry to create a hierarchy of relations so that changing the value of a parameter in the model (a length, an angle, an equation constant) would update all associated events in the tree and reconfigure the geometry of all related parts of the window. It took three months to reach the point where the design team was satisfied with the overall composition for the window. The exterior was fully resolved, while the interior had aspects that had to wait until certain measurements were made on site during construction before being finalised. Again, the use of parametric design technology allowed easy absorption of the new information as it became available, with a minimum of repeated work.

Prototyping

Three-dimensional wax printing was introduced as a means of rapid prototyping in order to overcome the distance between design and construction participants – the university team in Australia, the director and technical office on site and the stonemason in Galicia – and the unfamiliarity of team members with working directly from digital models without traditional gypsum maquettes to hold. This printing produced exquisitely detailed scaled versions of each individual stone in wax. The stonemasons developed their own, even more rapid, means to prototype, building full-scale contoured models in polystyrene sheet.

Six months into the project, while the lower quarter of the window was being constructed on site, the second quarter was still being cut in Galicia, the third quarter was made into templates in Australia to guide the stonemasons, and the design of the top quarter was being refined in collaboration between the Sagrada Familia design office and the team in Australia. These models were also a product of the digital process, constructed using sectional profiles (or surface contours) generated at regular intervals parallel to a specified Cartesian plane in the Rhinoceros software.

Template drawings

Nearly 800 full-size DIN A0 templates were mailed from Australia to guide the stonemason. Line colour was used to distinguish surface generatrices (green), template boundaries between adjacent pieces (black), curved surface intersections (orange), etc. The drawings were supported by surface contours at separations of 10cm, with coordinates at critical nodes linking the orthographic projections of each piece to a common datum.

Construction

Not only did the parametric software contribute to a flexible design process in which hard-to-attain measurements could be incorporated late in the process, it also contributed centroids and crane-lifting points to hoist the individual stones into position, perfectly orientated. Following the new system, all the A0 template drawings were employed immediately in Lugo without revision, and similarly the pieces were assembled for the first time on site, fitting together without the need for cutting or modification.

Project team

Architect coordinator and director: Jordi Bonet, Temple Sagrada Familia, Barcelona; project architect: Jordi Faulí, Temple Sagrada Familia, Barcelona; design and documentation: Mark Burry and Jane Burry, SIAL, RMIT University, Melbourne, Australia; project management: Ramon Espell; stonemasons: Manuel Malló Malló, Talleres Malló

The investigation of Gaudí's final design models for the Sagrada Familia church was supported by a Discovery grant from the Australian Research Council 2000-2002.

This work is supported by the Junta Constructora de la Sagrada Familia, Barcelona

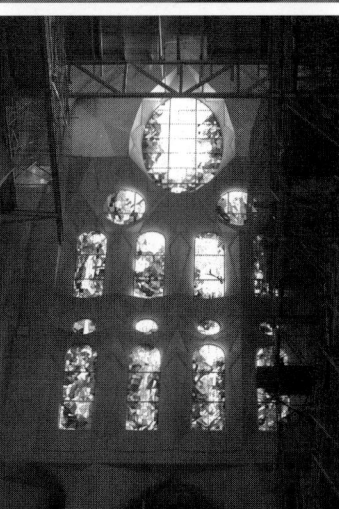

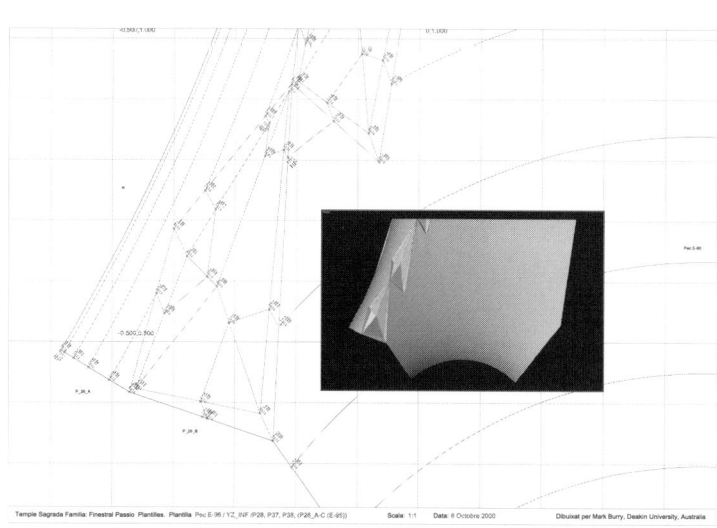

Temple Sagrada Familia: Finestral Passio Plantilles. Plantilla Pec E-96 (YZ_INF (P28, P37, P38, (P28_A-C (E-95))   Scala: 1:1   Data: 6 Octobre 2000   Dibuixat per Mark Burry, Deakin University, Australia

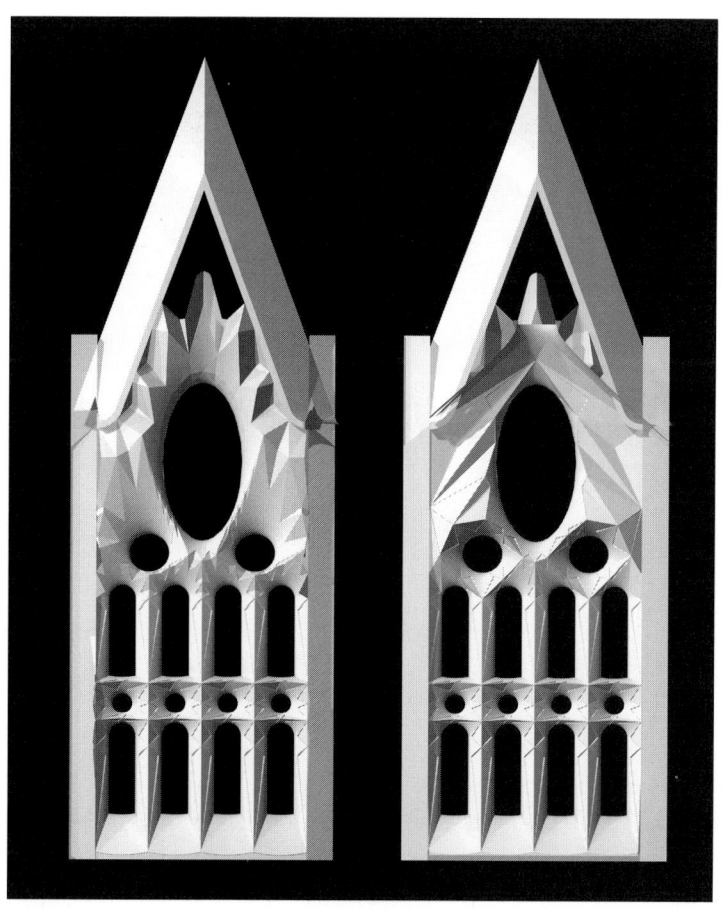

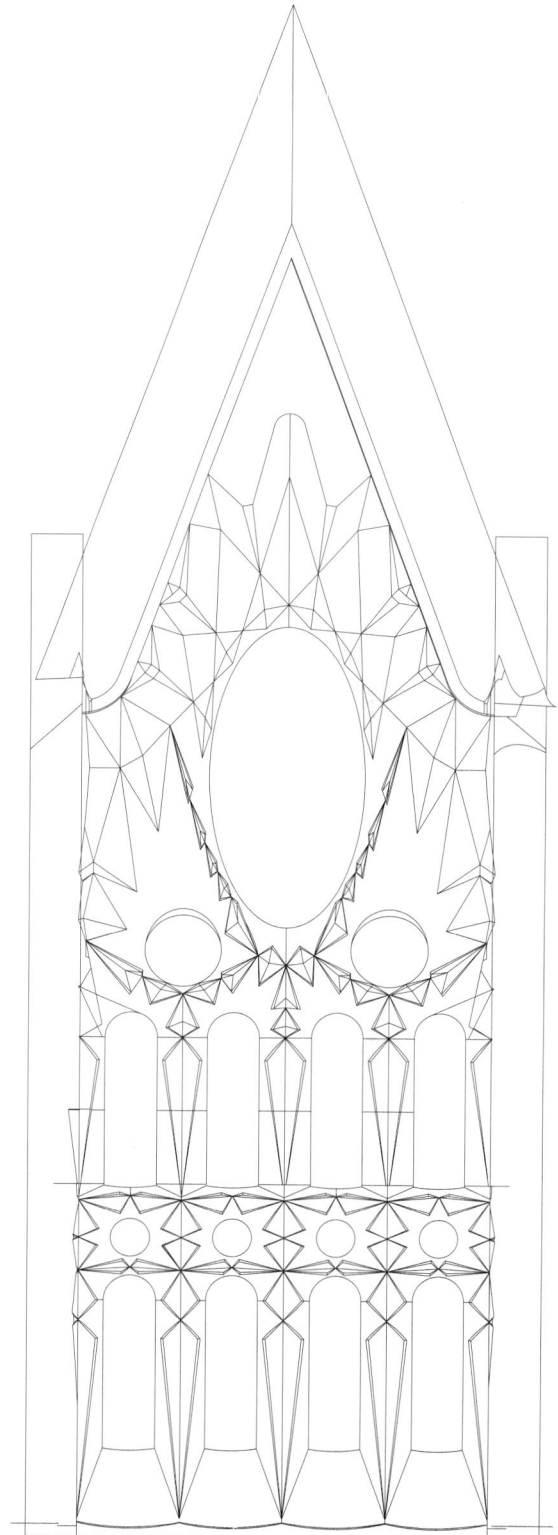

Swiss Re

Foster and Partners

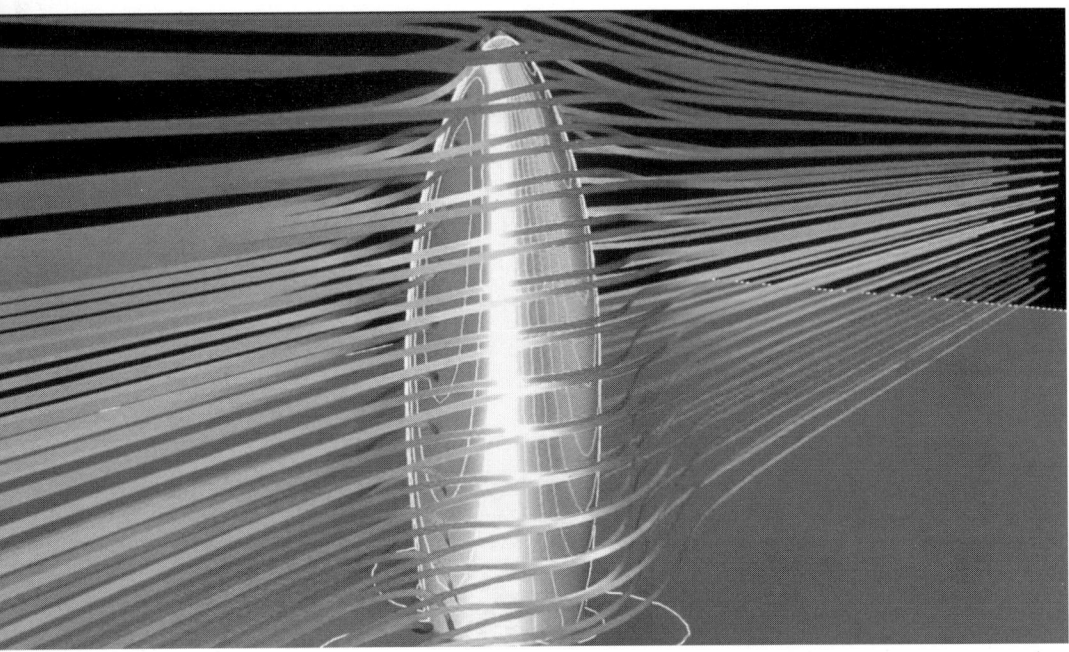

Foster and Partners' design for Swiss Re draws on advances in digital technologies to develop ideas first explored in the Climatroffice design with Buckminster Fuller in the early 1970s.

Swiss Re is on the site of the former Baltic Exchange in the City of London. The distinctive form of the 40-storey tower adds to the cluster of tall buildings that symbolise the heart of London's financial centre. It is the capital's first environmentally progressive tall building, and it is not only offices; its street level will be publicly accessible, with double-height retail outlets to serve the local working community, and the building is set within a new public plaza.

At the top of the building are dining and events facilities for the building's occupants and their guests. Beneath the glazed dome, a restaurant offers spectacular westerly views. The restaurant's mezzanine – a flexible space for drinks, gatherings or presentations – has a full 360 panorama over the city and beyond. The building is radical – technically, architecturally, socially, and spatially. Both from the outside and from within, it is unlike any office building so far conceived.

Aerodynamic form

The tower has a circular plan that widens as it rises from the ground and then tapers towards its apex. This form responds to the specific demands of the small site. The building appears less bulky than a conventional rectangular block of equivalent floor area; the slimming of the building's profile at its base reduces reflections, improves transparency and increases daylight penetration at ground level. Mid-height, the floor plates offer larger areas of office accommodation; the tapering peak of the tower minimises the extent of reflected sky.

The aerodynamic form encourages wind to flow around its face, minimising wind loads on the structure and cladding, enabling the use of a more efficient structure. Wind deflection to ground level is reduced, helping to maintain pedestrian comfort and safety at the base of the building. Wind tunnel tests have shown that the building will improve wind conditions in the vicinity. Natural air movement around the building generates substantial pressure differences across its face, which can be used to facilitate natural ventilation within the building.

Parametric modelling

Conceptually the project develops ideas first explored in the Climatroffice design with Buckminster Fuller in the early 1970s. That project envisioned office space enclosed within a free-form glass skin to create a building with its own microclimate. At the time its complex, double-curved geometry would have been difficult to build. Thirty years later, digital technologies facilitate the design and construction of buildings such as Swiss Re in a fraction of the time it would have taken in the 1970s.

Parametric modelling, originally developed in the aerospace and automotive industries for designing complex curved forms, has had a fundamental effect on the design of the building. The three-dimensional computer modelling process works like a conventional numerical spreadsheet. By storing the relationships between the various features of the design and treating these relationships like mathematical equations, it

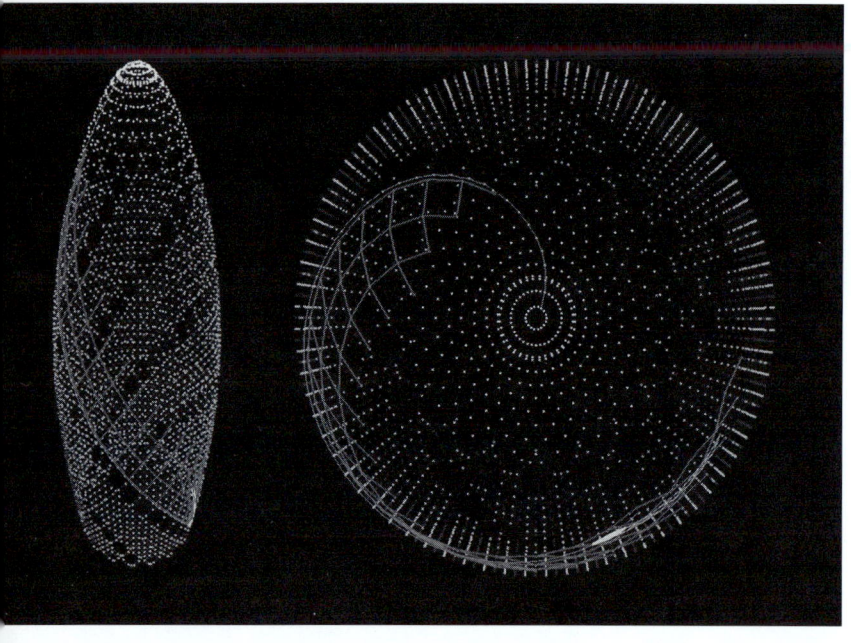

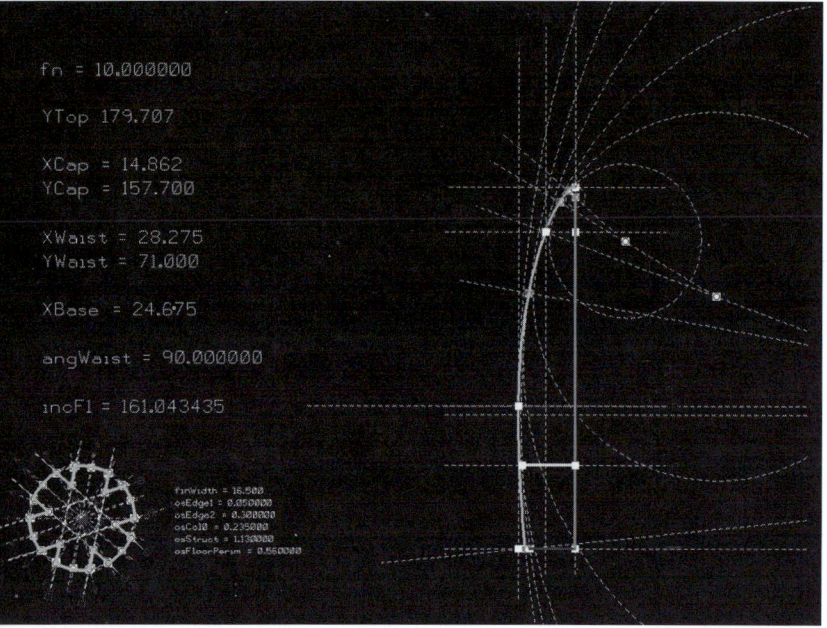

allows any element of the model to be changed and automatically regenerates the model in much the same way that a spreadsheet automatically recalculates any numerical changes. As such, it becomes a 'living' model – one that is constantly responsive to change – offering a degree of flexibility never previously available. The same technology also allows curved surfaces to be rationalised into flat panels, demystifying the structure and building components of highly complex geometric forms, allowing them to be built economically and efficiently.

Spiralling lightwells

Swiss Re can use natural ventilation in addition to air conditioning, so that for up to 40 per cent of the year much of the mechanical cooling and ventilation supply systems can be supplemented, reducing energy consumption and carbon dioxide emissions. Fresh air may be drawn through lightwells, which spiral up the building, to ventilate the offices naturally. The building's aerodynamic form generates pressure differentials on the facade that greatly assist this natural flow.

The lightwells maximise daylight penetration to the office accommodation and therefore reduce reliance on artificial lighting. They also help to break down the scale of the floor plates, while externally they add variety and life to its facades. The balconies on the edge of each lightwell have strong visual links between floors and provide a natural focus for communal office facilities such as refreshment points, copy centres, filing and impromptu meeting areas.

Public space

If the building occupied the whole of its site it would leave very narrow streets at its edges. Instead, the building has a relatively small circular footprint, allowing the remaining space at ground level to be landscaped for public use, and improving the surrounding environment. A new plaza is contained within low stone walls, which define the historical site boundary and act as elements of street furniture. High-quality materials, including natural stone, stainless steel and aluminium, are used throughout. Mature trees form an integral part of the public realm.

Structure and cladding

The 180m-tall tower is supported by a highly efficient structure consisting of a central core and a perimeter diagrid – a grid of diagonally interlocking steel elements. Some traditional central-cored buildings of this height would use the core as a means of providing the necessary lateral structural stability. Because of the inherent stiffness of the external diagrid, the central core is required to act only as a load-bearing element and is free from diagonal bracing, producing more flexible floor plates.

The fully glazed skin of the building allows the occupants to enjoy increased external awareness and the benefits of daylight. The glazing of the office areas comprises two layers of glass with a cavity, which is ventilated by the used air drawn from the offices. This enables solar radiation to be intercepted before it reaches the office spaces to reduce the typically large air-conditioning load. The cladding of the lightwells consists of simple operable and fixed double-glazed panels, with tinted glass and a high-performance coating to reduce the penetration of solar radiation.

Architect
Foster and Partners

Structural Engineer
Arup

Main Contractor
Skanska Construction UK
Mechanical & Electrical Engineer
Hilson Moran Partnership

Fire Engineer
Arup Fire

Cladding Consultant
Emmer Pfenninger

Lighting
Speirs and Major

Acoustics
Sandy Brown Associates
FaCade Access
Reef UK

Landscape Architect
Derek Lovejoy Partnership

Information Technology
PTS

Planning Supervisor
Osprey Mott MacDonald

Project Management
RWG Associates

Cost Consultant
Gardiner & Theobald

Planning Consultant
Montagu Evans

Environmental Engineer
BDSP Partnership

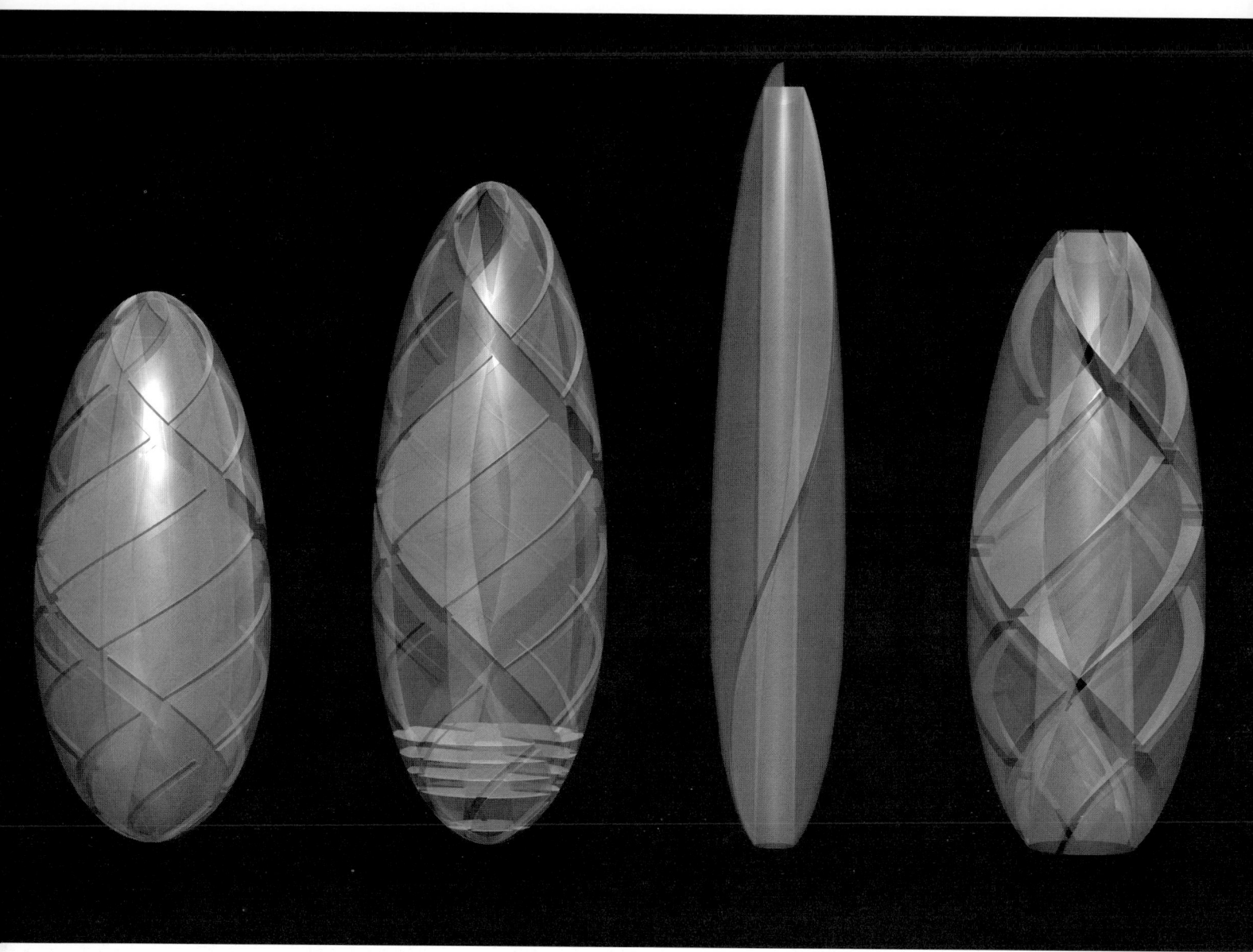

Selfridges Link Bridge

Arup with Future Systems

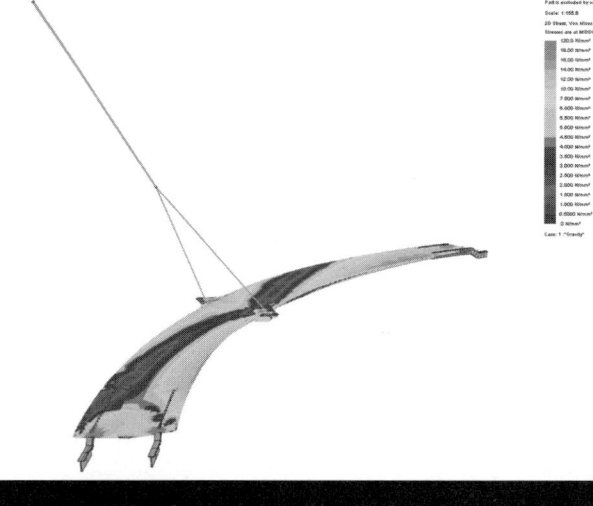

Orpheus Filter

Philip Beesley with Diane Willow and Steve Wood

For the *Digital Fabricators* exhibition, the Canadian architect Philip Beesley has produced a prototype building membrane. Orpheus Filter is conceived as an 'interactive geotextile' and is part of a series entitled *Reflexive Membrane* being developed with artist and scientist Diane Willow and systems designer Steven Wood. The fabric comprises a dense array of interlinking elements, making an intricate three-dimensional structure. It is equipped with layers of miniature valves and clamping mechanisms that slowly digest and convert surrounding material into a fertile living wall. The project draws on Beesley's knowledge of industrial design and architectural textiles (honed at the University of Waterloo's Integrated Centre for Visualization, Design and Manufacturing) and on the expertise of Willow, who specialises in integrating flexible miniature computer components (microprocessors, sensors, actuators) at the MIT Media Lab.

The work is a speculation of about building living skins. In 2004 projects in Birmingham and London, and most recently in Cambridge, near Toronto, Canada, lightweight expanded meshworks are installed within rooms making an immersive lining. The structure acts like an artificial reef that could support a turf-like surface of natural material. The structure in these spaces respond to the viewer, hovering and vibrating in response to currents and pressure. The project probes the possibilities of combining artificial and natural processes to form a hybrid ecology.

Orgone Reef is in part a technical exercise in construction and fabrication. The project relates to geotextiles, a new class of materials used for reinforcing landscapes and buildings. A minimal amount of raw material is expanded to form a network forming a large, porous volume. A Penrose tessellation, a non-repeating geometrical system, is used to organize the interlinking elements of this hybrid fabric. The installations are dense matrices made of thousands of pieces manufactured using automated laser cutters working directly from digital models. Individual elements can be produced at low cost and quick cycles of refinement using this method, supporting highly efficient industrial design. The incremental scales of the production suggests a cottage-industry based economy.

At the same time, the projects tend to question boundaries of psyche. Their large-scale field structures offer immersion, an expansion rendering our bodies porous and offering wide-flung dispersal of identity. The title *Orgone* is derived from a term coined by Wilhelm Reich, a psychologist working alongside Freud, that describes a subtle life force encircling the world. Reich, whose work was tinged by obsession, saw the world as an intelligent, evolving entity. His visions offer a poignant alternative to the Modern version of progress.

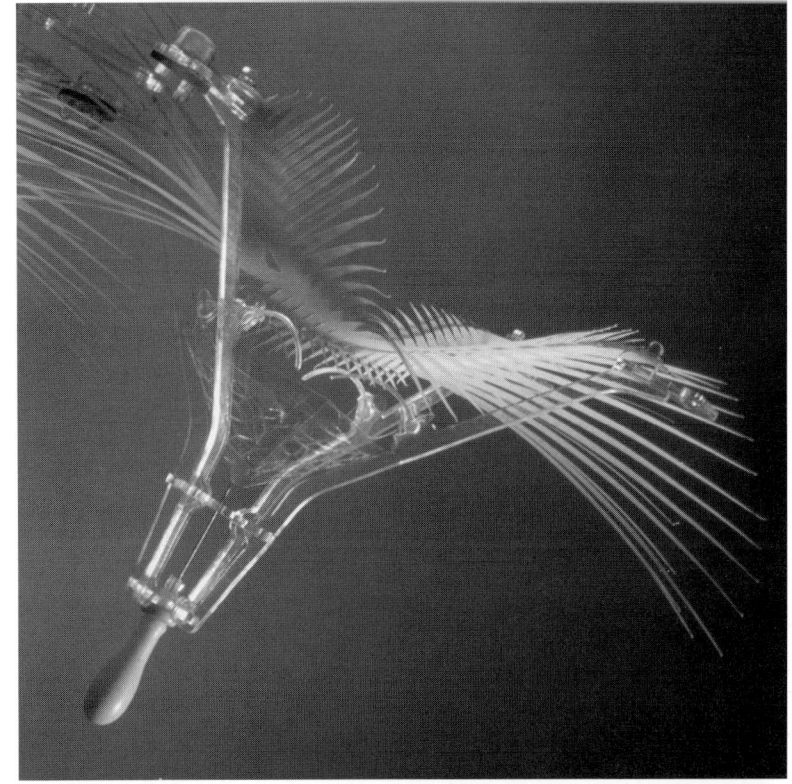

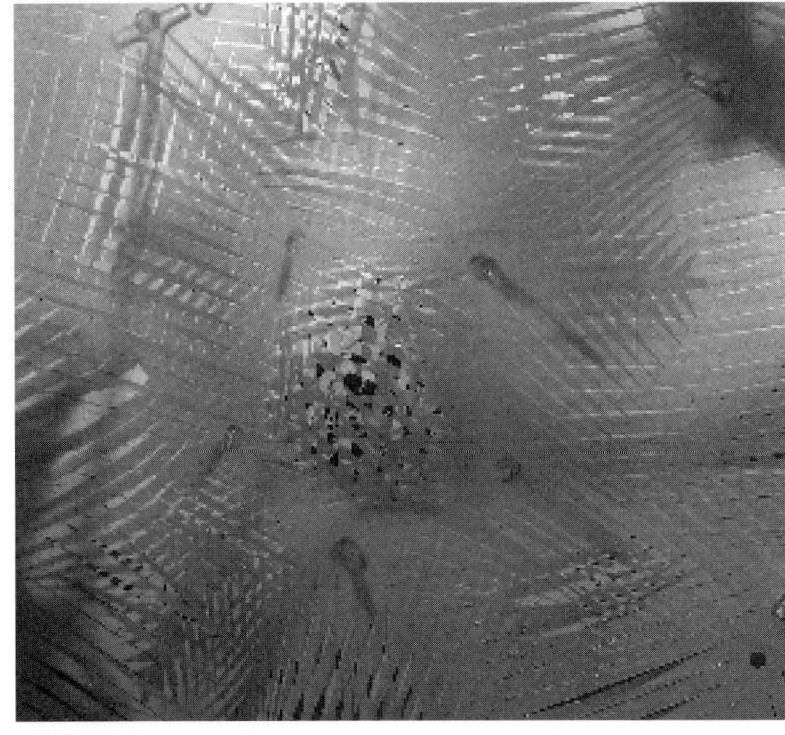

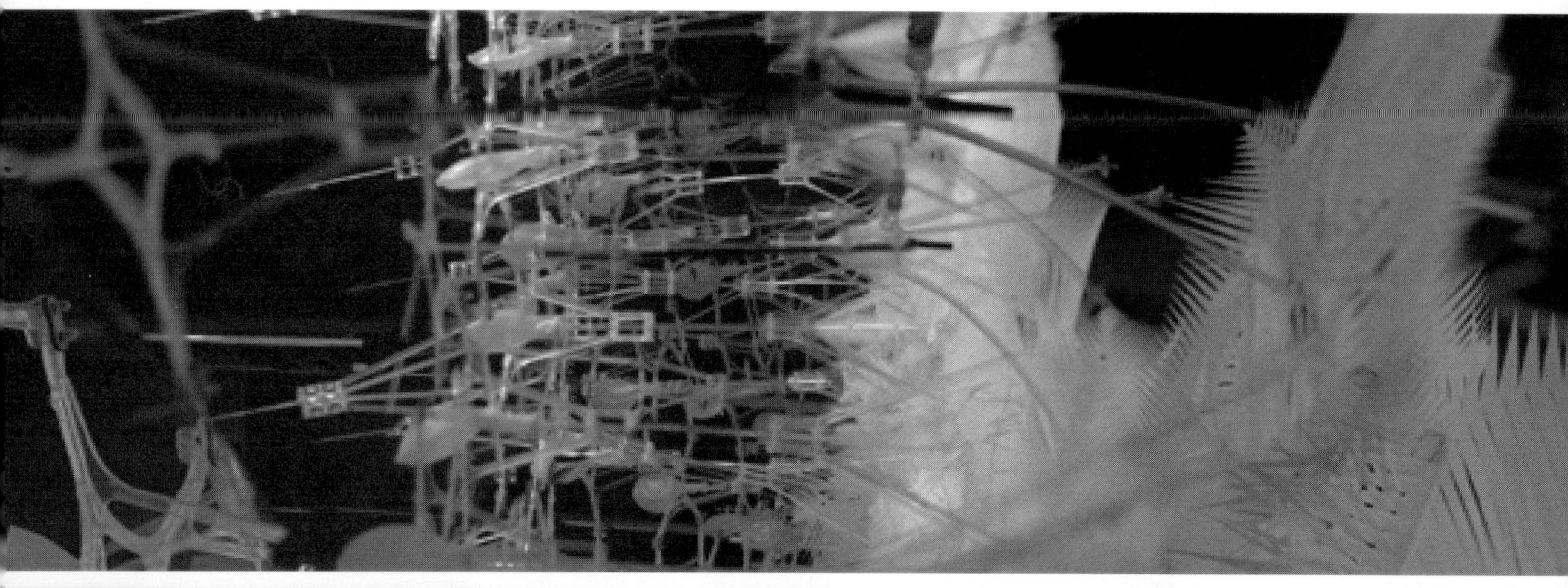

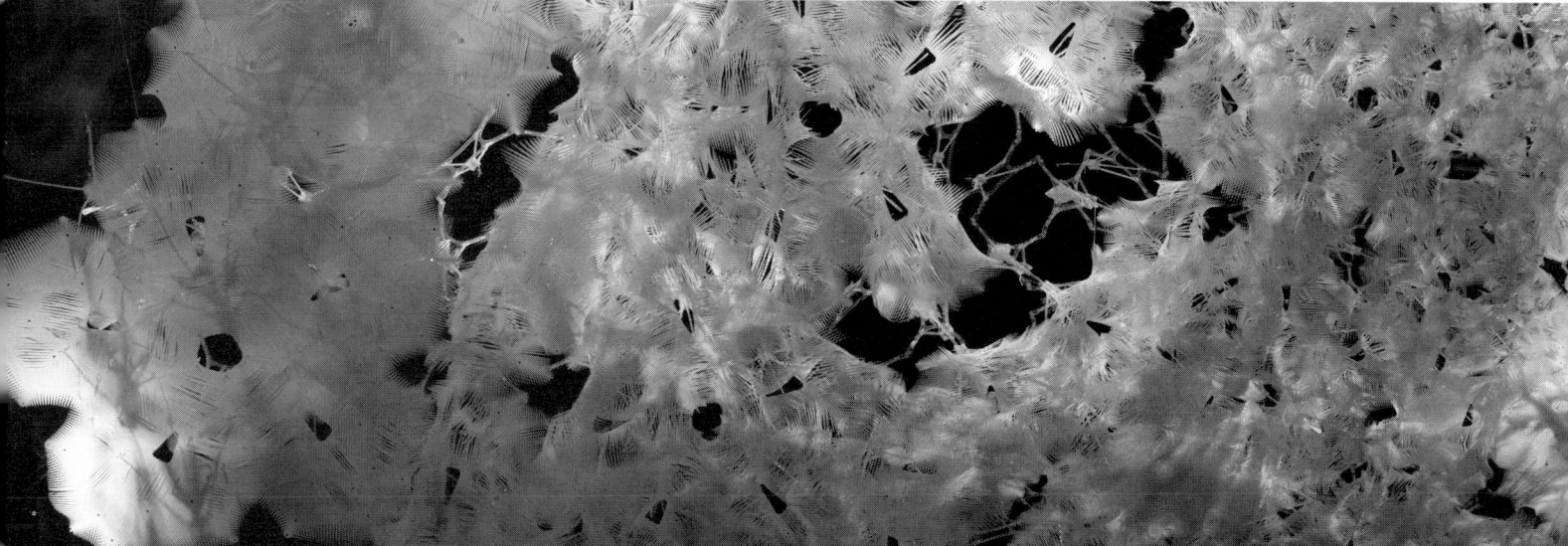

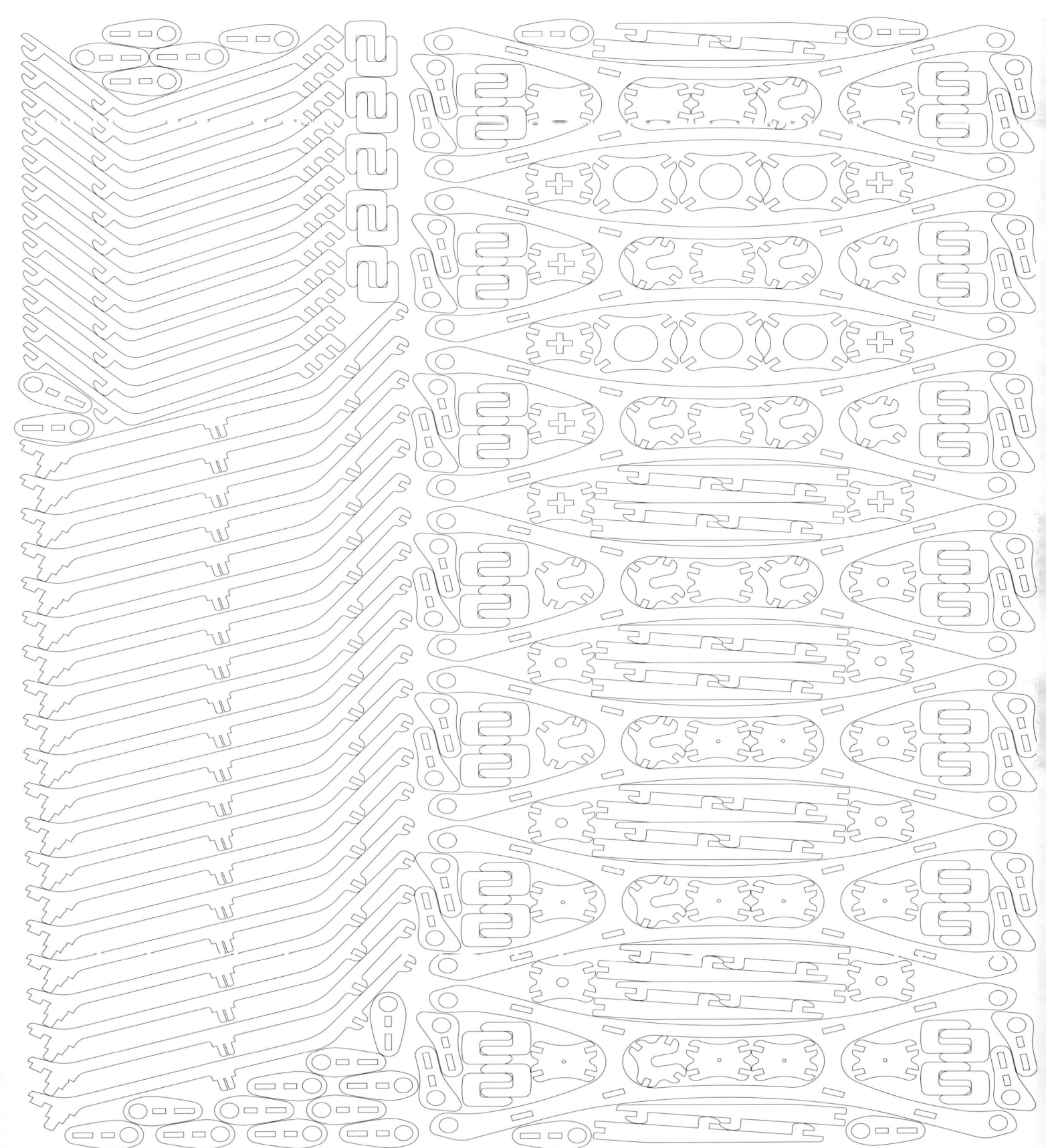

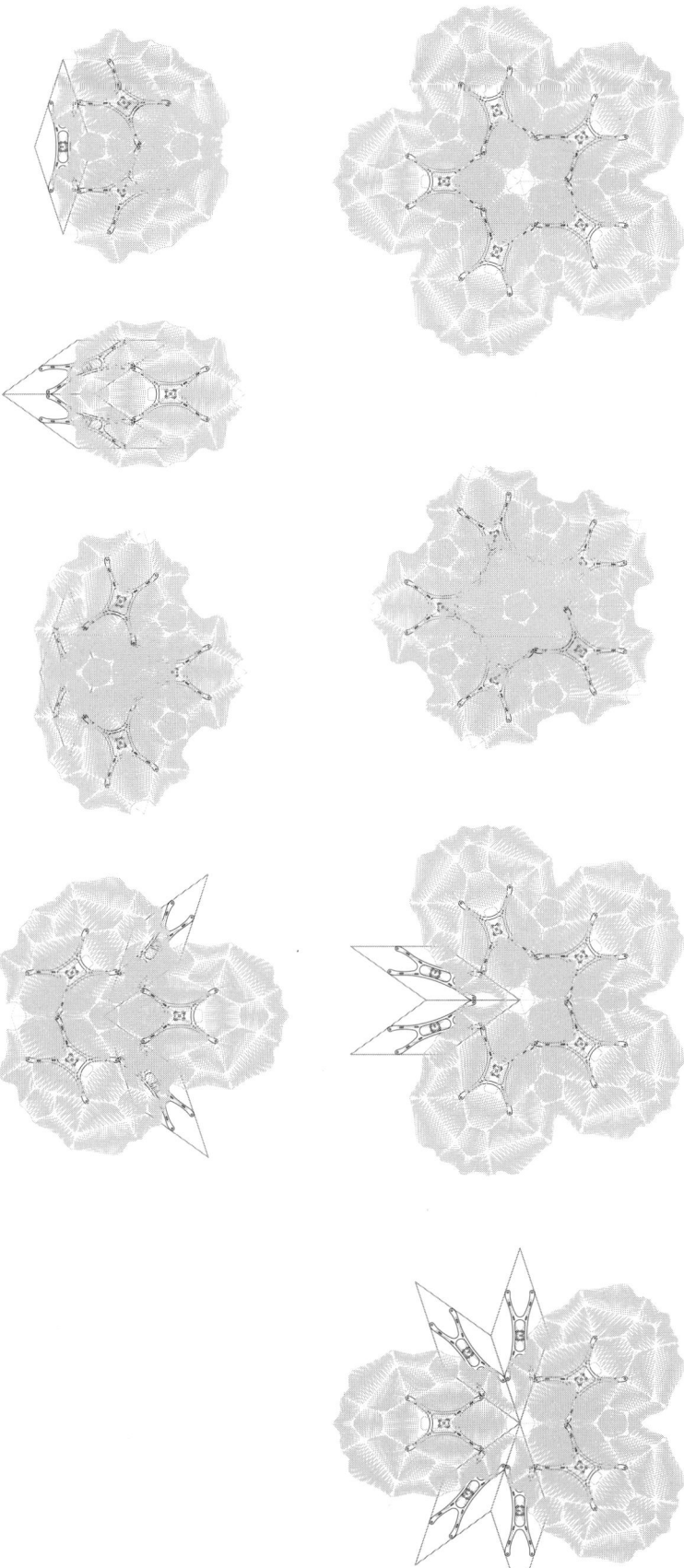

Credits
Philip Beesley, University of Waterloo

Consultant
Diane Willow, MIT Media Lab

Computing Systems Design
Steven Wood

Design Assistants
    Jonathan Tyrrell
    Coryn Kempster
    Farid Noufailly
    Will Elsworthy
    Alex Josephson
    Luanna Lam
    Thomas-Bernard Kenniff
    Alex Lukachko
    Vincent Hui
    Kirsten Douglas
    Daniel Hall
    Nancy Gibson

Prototyping

University of Waterloo
Integrated Centre for Visualization, Design and Manufacturing

Ballingdon Bridge

Brookes Stacey Randall Architects

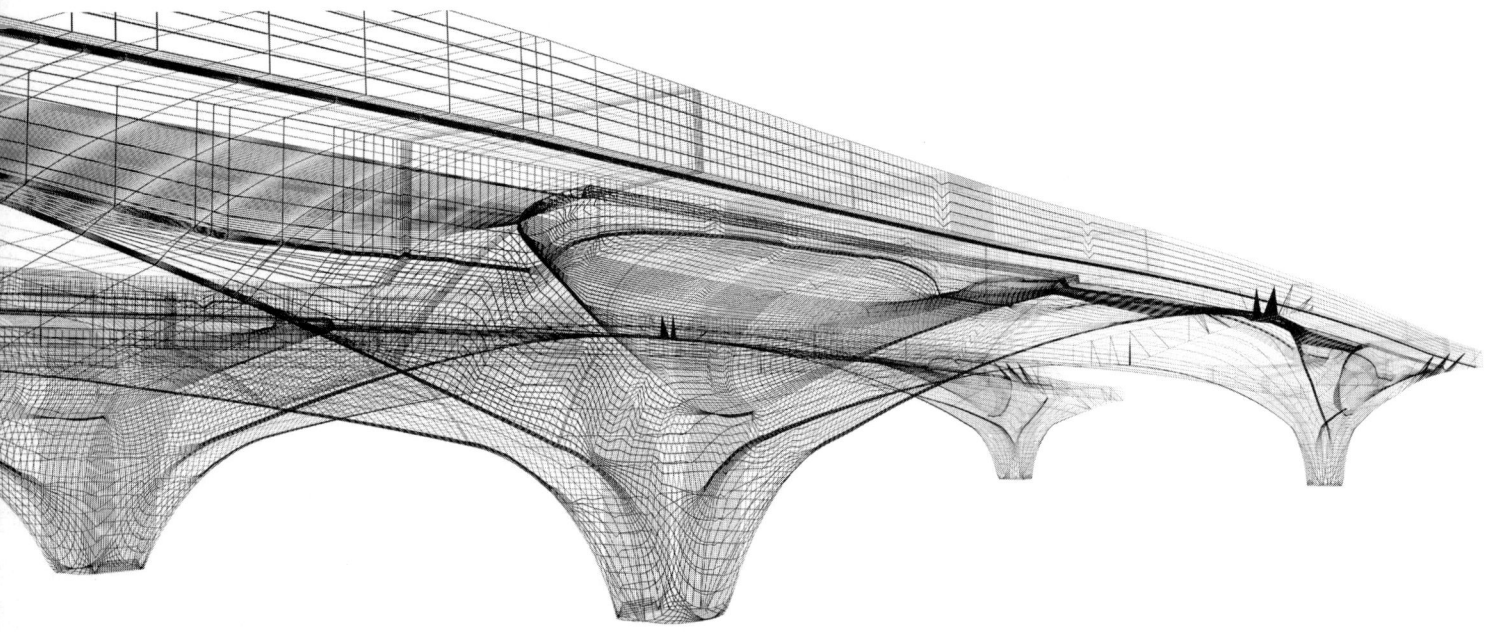

Brookes Stacey Randall Architects used precast units crafted from digital geometry to construct an elegant road bridge over the River Stour.

The new Ballingdon Bridge, designed by Brookes Stacey Randall and Arup, is an integral reinforced concrete structure that carries the A131 over the River Stour, replacing the existing bridge, which can no longer sustain 42-tonne articulated lorries. The setting is a wonderful combination of the water meadows that surround Sudbury and the listed buildings that form a conservation area, including the village of Ballingdon. There have been bridges on this site since the 12th century. The previous bridge, built in 1911, was not capable of sustaining heavy traffic and closure would have resulted in a 35-mile diversion.

The design of the new bridge is visually calm, respecting the historic context; however, the structure has a dynamically three-dimensional soffit. Designed using an evolutionary technique, the bridge has an every-change and site-specific geometry. The feet of the new bridge appear to touch the surface of the water lightly. The primary structure of the bridge is formed from precast concrete; the mix has been selected to match the local limestone of the Norman church. The precast units were manufactured by Buchan in timber moulds, which were beautifully crafted from Brookes Stacey Randall's digital geometry. By careful study of the construction and phasing of the bridge, and extensive prefabrication, disruption to Ballingdon was minimised and two-way traffic on the bridge was maximised during reconstruction. The bridge was rebuilt in 18 months and has a design life of 120 years. Brookes Stacey Randall sought to uphold the rich architectural traditions and construction quality of Suffolk – Sudbury was the home of Gainsborough, and the landscape of the river Stour is Constable country.

The balustrade has been designed to be visually open so that the views of the landscape are as uninterrupted as possible. This 'P2 Low Containment' balustrade, in combination with the bollards, is capable of arresting a 42-tonne truck, yet appears to be an elegant pedestrian handrail. The traffic and pedestrian functions of the bridge have been safely separated, the pavement being protected by bollards that also house light fittings. The enlarged walkways create a generous provision for pedestrians to appreciate the views of the river and meadows. People enjoying the river and the urban spaces of Ballingdon and Sudbury are the priority within the design of this road bridge.

Credits

Project Team

Brookes Stacey Randall Architects: Michael Stacey (partner in charge), Laura Irving (project architect); Ove Arup and Partners: Ian Burton and Stuart Smith; Suffolk Highways Engineering Consultancy: Andy Bilby and Chris Graves; Equation Lighting Design: Helen Milligan and Alexis Themis

Client/Sponsor

Suffolk County Council, Department of Environment and Transport.

Bridge improvement and Maintenance Manager

Nigel Burrows

Contractor

Costain

Other participants

Suffolk County Council Planning Department, The Landscape Partnership, and the Environment Agency. Delivering the design has taken close collaboration and teamwork, facilitated by partnering

Photography

By the design team including: Andy Bilby, Laura Irving and Michael Stacey

Seattle Central Library

FRONT Inc.

Front Inc. working in association with Dewhurst Macfarlane & Partners served as façade engineering consultants for the new Seattle Central Library project. Front worked intimately with the complete project team in a highly collaborative environment including the OMA/LMN Joint Venture (Architects), Magnusson Klemencic/Arup (Structure), Arup (Services), Hoffman Construction (GCCM) and Seele (Façade Sub-Contractor). Front joined the project team at the end of schematic design and continued through to the completion of the installed façade works.

During the design process the architecture evolved through several highly distinct iterations including a building wholly clad in engineered insulated fabric. Driven by a strong desire to unify and simplify the design given the complex geometry; and by a demanding value-engineering process, a design solution emerged that integrated the structural and façade geometries. The sloped structural diamond-grid steel performs several functions and through many studies a grid density was developed that could support a fine façade system while transfering both wind and seismic loads. This approach elegantly reinforced central themes to the building and resulted in signifcant cost savings essential to maintaining the project budget.

The design team worked to design a single façade system concept that could be universally applied to the complete building enclosure. The final configuration achieves this intent through a design that retains uniform system integrity through the first several layers of the façade. The system is then modified and adpated in every regard to accommodate the performance specificity required in each distinct condition, resulting in a fundamentally modular component-based façade system that relies on full pre-engineering and numerically controlled pre-fabrication but with the majority of elements assembled on-site. A consensus was reached by both design and contractor teams that this particular combination of comprehensive component pre-fabrication and site assembly was essential to the successful resolution of a myriad of design, geometry, performance and installation requirements.

The design employs fabrication techniques applied to the diversity of materials used in the system. Some examples are the following;

NC laser cutting and water-jet cutting/indexing of the integrated metal mesh and glass panels of which there is 135,000sf of material with 65 percent of panels being of non-standard geometry.

Casting and machining of the stainless steel façade access tie-back anchors.

Casting of the POM (commonly used automotive plastic) bearing blocks and compression spacers.

Extruding, NC-milling and stamping of anodized aluminum bracket assemblies.

NC-milling of custom stainless steel hardware fixing elements.

Extruding, and extensive NC-cutting, drilling and milling of custom anodized aluminum façade profiles.

Application of innovative flexible aluminum-reinforced butyl tape as primary water-proofing seal to accommodate seismic racking.

Extruding and injection-mould casting of silicone-based gasket materials.

NC-cut, break-formed and welded custom anodized aluminum sheet corner and gutter assemblies.

Computer controlled breakforming of stainless steel support channels for butyl tape.

NC-cutting and milling of stainless steel snow and ice guards.

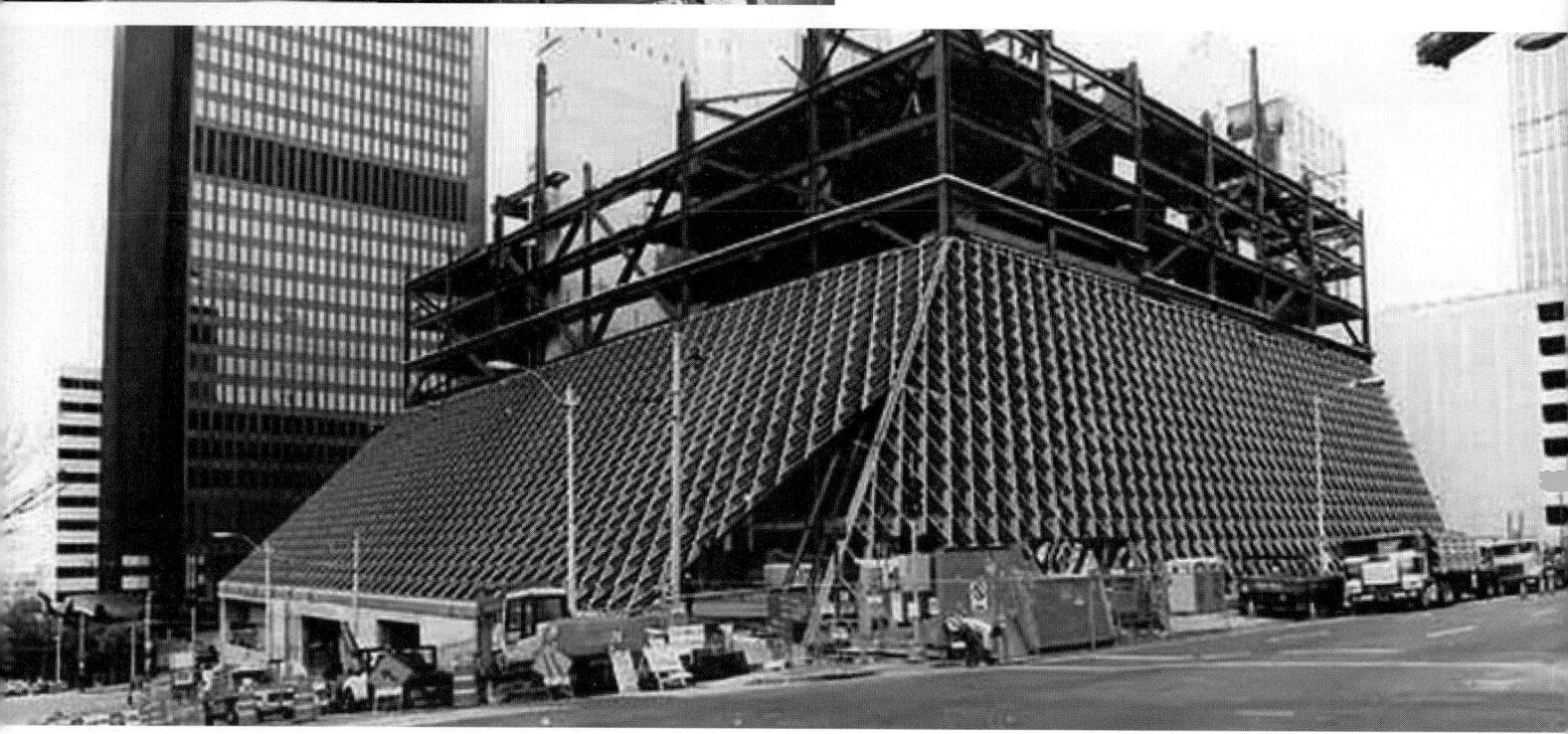

SCL Glass Office & Showroom

FRONT Inc.

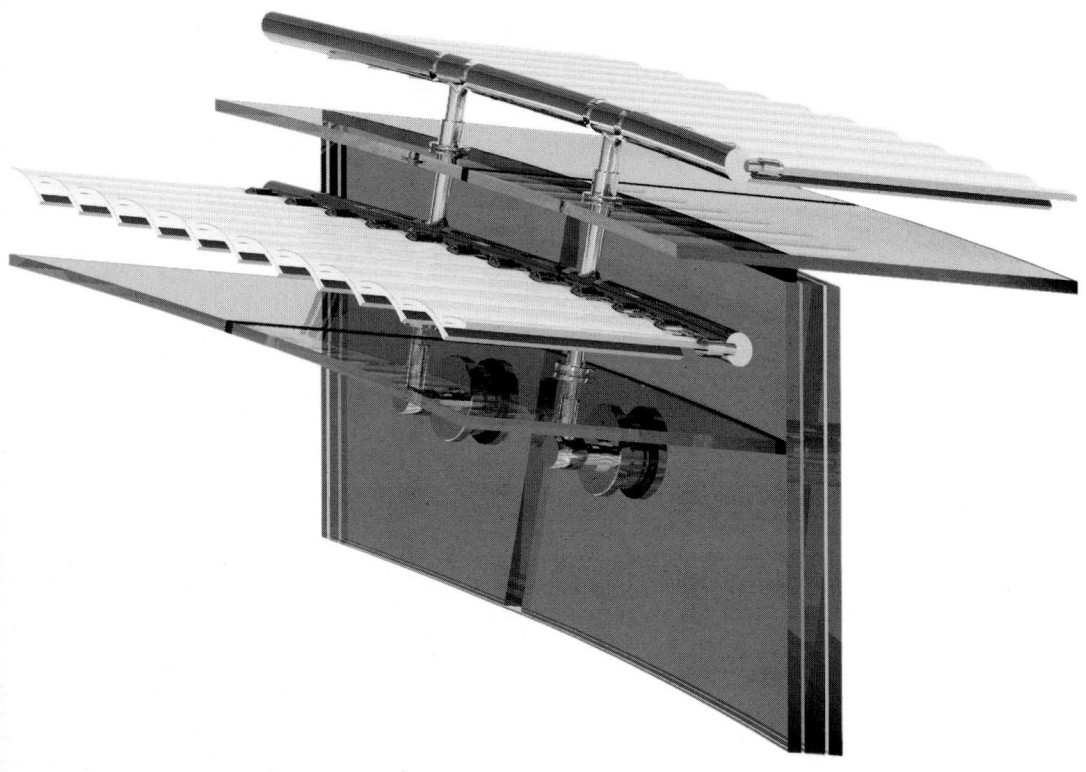

Front Inc. are Design Architects, Façade Engineers and Structural Engineers of Record for the new SCL Glass & Office Showroom project in Yatala, near Brisbane, Australia.

The project is an office and showroom adjacent to a glass manufacturing plant that is currently under construction in Queensland, Australia. The client required a design vocabulary that would optimally demonstrate the production capabilities of the factory. The resultant design features a 70ft × 130ft continuous enclosure composed of overlapping, curved ribs of toughened, laminated glass mechanically interlocked with steel pins set with a temperature stable pressure injected grout. All the panels of the building envelope are composed of multiple layers of glass, fabricated in the adjacent factory, directly from digital shop drawings produced using Gehry Technologies' CATIA/Digital Project.

The design was generated from an elliptical form which was then refined to meet the client's manufacturing capabilities. Working with CATIA/ Digital Project, the ellipses were transformed into a radial geometry that is factory and material-friendly. Further refinement involved an iterative process of re-generation of the rib forms to directly express the inherent structural forces. This resulted in a glass rib form that reflects the bending moment diagram, maximising the efficiency of the structure. The roughly 1200 shop drawings required to fabricate the glass are produced seamlessly from the digital model. This high degree of automation manifests itself in the overall Architectural expression of the finished project. Despite the enclosure's dramatic shape, the glass has no double curves.

Star and Cloud

Bruce Gernand

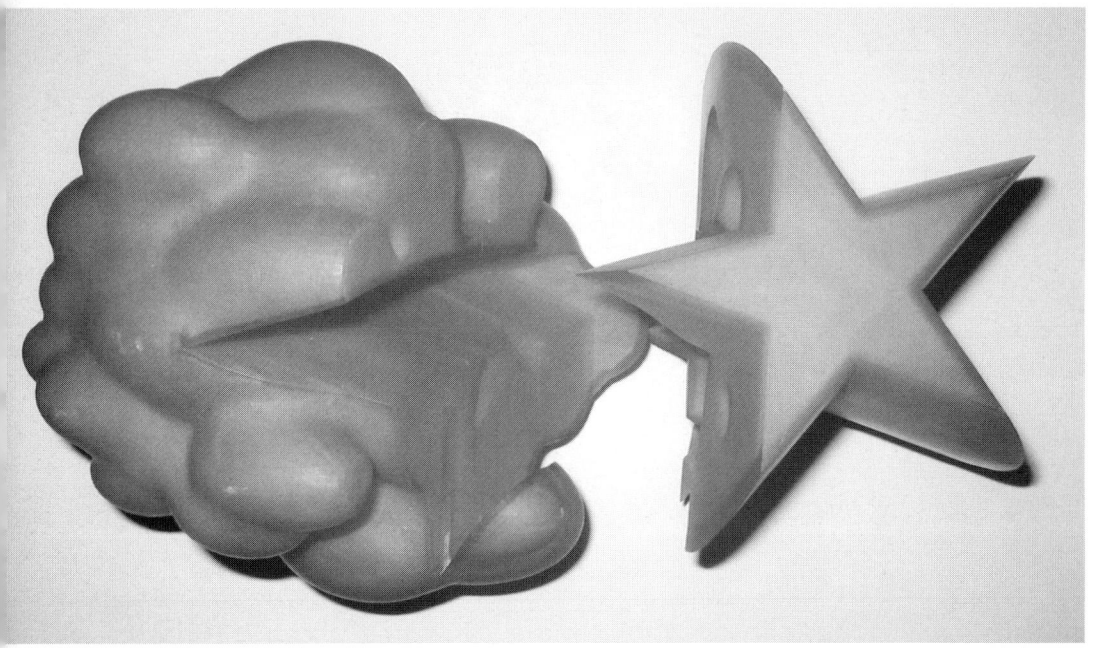

Bruce Gernand is a sculptor whose work questions the relationship between the virtual image and the firm reality of the sculpted form.

Star and Cloud, commissioned by Sculpture at Goodwood was his first major investigation of using a CAD programme (Form Z) to enable the contours of one object to remove areas of another object. This Boolean intersection was employed in a simultaneous way – each object intrudes on the other's shape – so that a section of the 'cloud' shape exists as an impression in the 'star' shape. By reading the 'negative' surface of one object the other object's 'absent' shape is revealed.

Gernand also utilises the computer for both outputting (e.g. rapid prototyping) but also to produce models in real space through both contour slicing (where the computer makes a series of regular slices through an object which are then printed and used as cutting templates) and to unfold an object (where the computer model is 'unfolded' to produce a two dimensional pattern of interconnected facets which can be printed and glued to card to produce a 3D model).

Bahá'í Temple for South America

Siamak Hariri

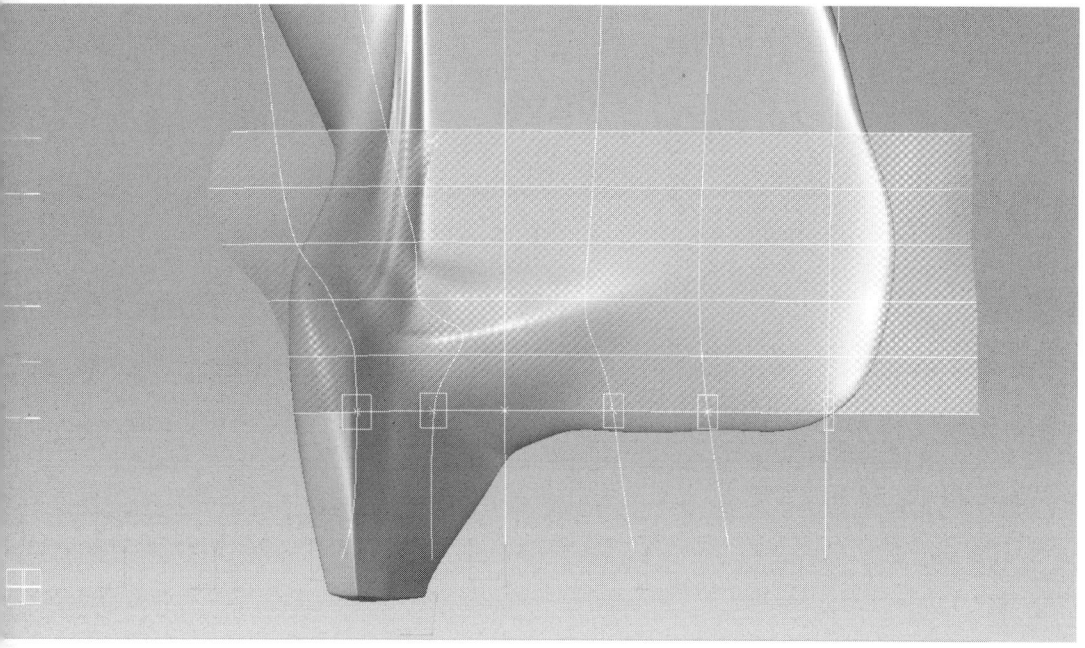

Born from an international competition to design the Mother Temple of South America –the last of seven continental temples for the Bahá'í people – our design conforms to the simple requirements of all Bahá'í Temples: a domed structure with nine sides and nine entrances to symbolically welcome all people from all directions of the earth for prayer and meditation.

Employing light as inspiration, the design team looked to shapes which allow for the passage of light in complex and intricate patterns: blossoms, swirling skirts, frozen icicles, hand woven baskets, snail shells, fruit, vegetables and the human heart.

Set against the dramatic Andes Mountains, the resulting Temple of Light 'floats' over nine reflecting lily pools and nine prayer gardens. Nine gracefully torqued wings of translucent stone billow like sails, projecting a sublime, ethereal luminescence.

Inside, the dome shields a filigree basket of white oak which dapples the interior walls with dancing patterns of light and shadow. Seating 500 people, the open interior is brought to a human scale with a mezzanine ringing the perimeter, and alcoves set within the wings for private meditation. Windows through the partings of the wings, guide streams of light in, and provide views out to the gardens and the surrounding mountains.

If visually light, the building is structurally strong. Situated in a seismic zone, the structure is designed, as much as technically possible, to accommodate ground movement as well as flex under lateral loads. Each wing is composed like a leaf, whose veins stem from a primary steel structure with secondary branches, thinly veiled, supporting the external skin of stone, and an internal skin of white oak. These are tied together by three rings.

It is hoped that this sacred building will feel both simple and understated, and, at the same time, complex enough to accept and hold a rich multiplicity of readings and experiences.

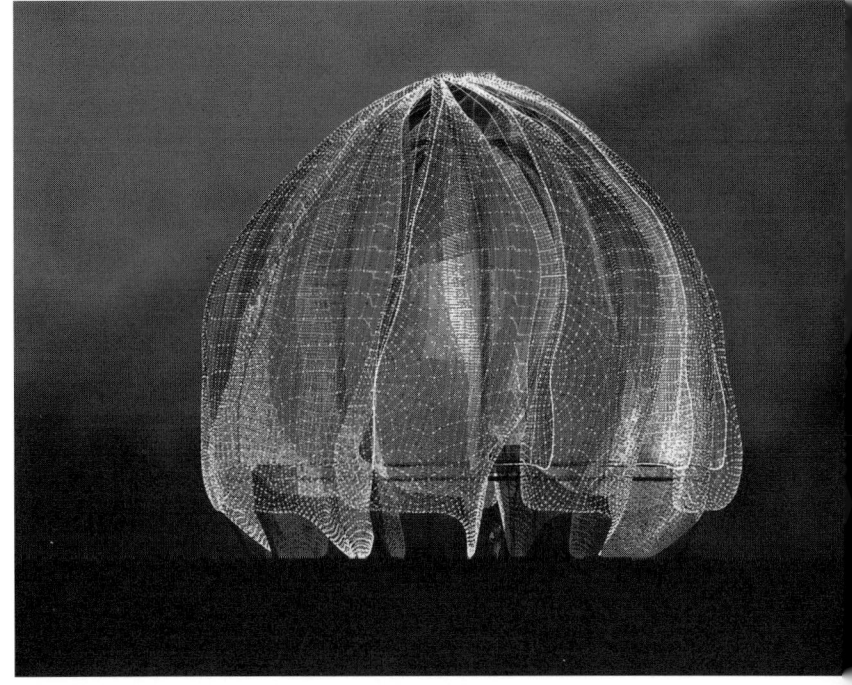

The National Stadium, Beijing

Herzog and de Meuron and ArupSport

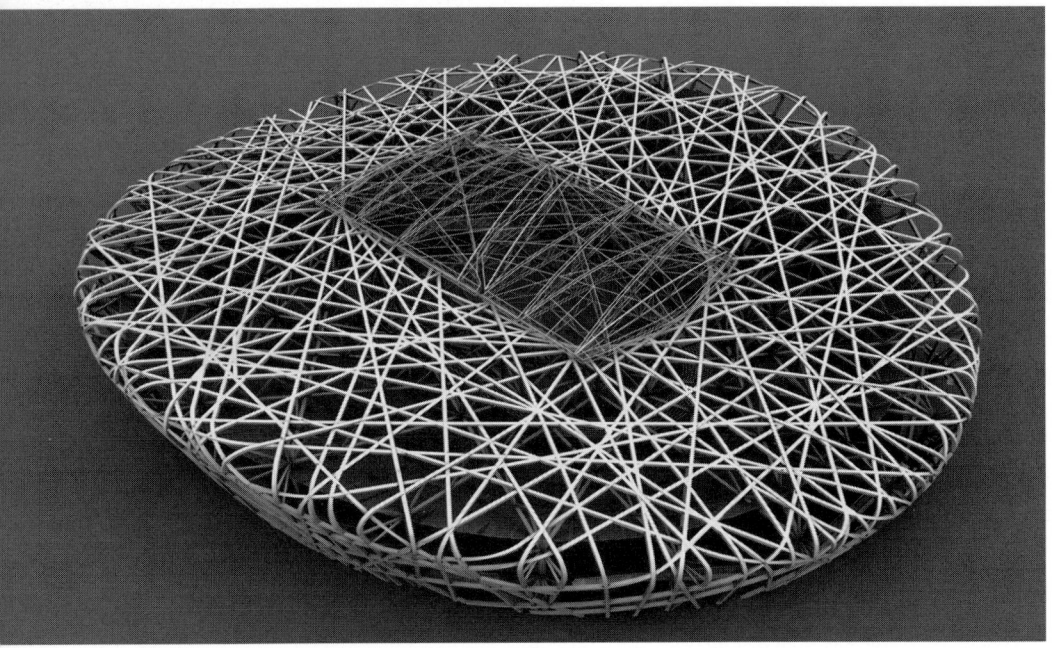

The exploration of the complex form and geometry of the Olympic stadium in Beijing by Herzog & de Meuron and ArupSport was made possible by a range of tools including CAD visualisations and rapid prototypes.

The main stadium for the XXIX Olympiad, to be held in Beijing in 2008, designed by a team of Herzog & de Meuron, ArupSport and China Architecture Research Group, will be an iconic form.

The concept behind the design is to create a form that is pure structure with an almost archaic quality. The elements for the facade and roof mutually support each other to create an integrated structure that blurs the line between primary and secondary elements. The apparent chaotic pattern is governed by a series of 'rules' that generates the nest-like structure.

Geometry

The pattern for the structure, in addition to forming an interwoven mesh of members, also follows a three-dimensional, undulating form that mirrors the profile of the seating bowl contained within. The constraints of seating 100,000 people round an athletics track and field create an overall elliptical base, and it is this parameter that determines the plan form of the structure. In cross-section the majority of the spectators are contained on the long sides of the track; at the ends of the stadium the seating tiers are smaller. This profile ensures that all the spectators are within the same radius of view from the corners of the field.

From these constraints a set of neutral surfaces can be generated for the inner and outer elements of the roof structure.

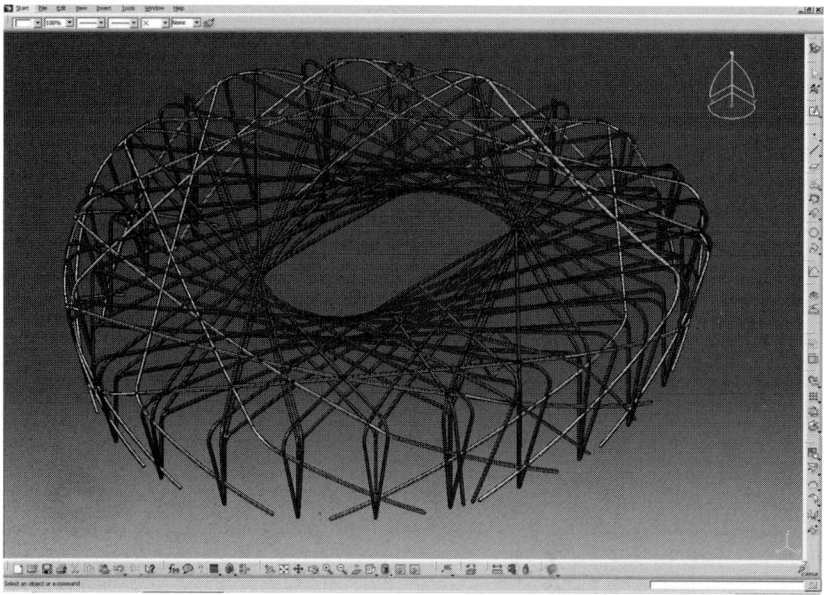

The outer surface is constructed from three principal surfaces:

- a toro id patch for the roof surface;

- a conical ellipse for the facade surface; and

- a radiused fillet between the toroid and the cone.

Having defined this surface, the structural members can be projected from their plan geometry on to the three-dimensional surface. The members are classified into a number of different categories, which respect their function within the structure as a whole:

- the primary structure, consisting of a series of tangential trusses that intersect to create a three-dimensional space frame structure. This is the only element of the structure that extends out of the neutral surface plane, to create a 12m-deep truss;

- the stair-geometry, facade elements defined by the perimeter circulation stairs. These are mapped on to the elevation and then extended across the roof surface;

- the radial infill pattern;

- additional infill members.

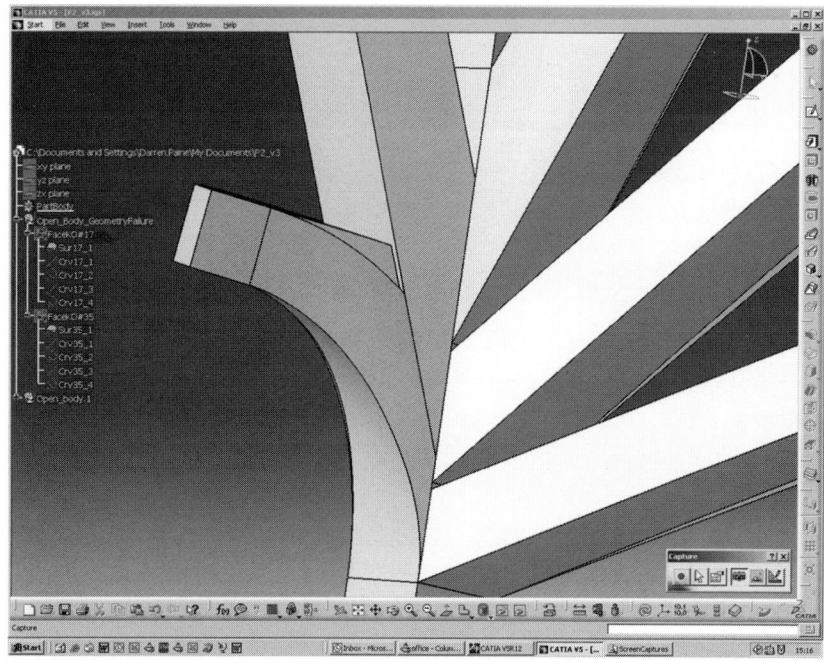

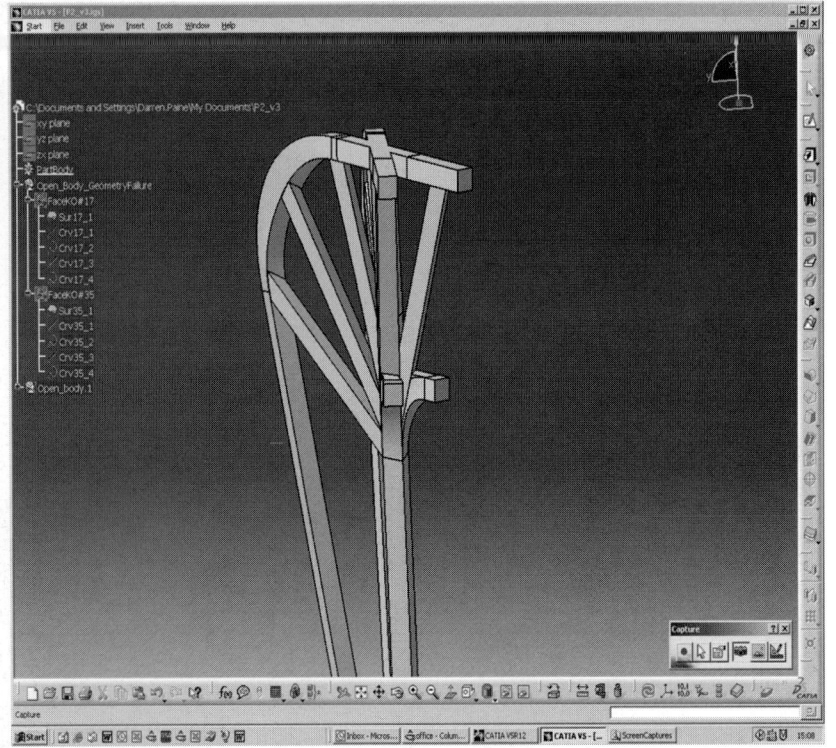

Each member of the above groups is formed from a square steel box section. These sections twist and rotate as they cross the surfaces to ensure that the top flange of the member is parallel to the surface in all instances.

Primary column

Once the base geometry has been created, the detail construction of principal components can begin. Illustrated overleaf is the detail at the top of the primary column where the trusses intersect. This is one of the major structural connections within the roof as it is the focus for the applied loads to be transferred from roof to foundation.

In addition to the digital studies, a rapid prototype of the model was also constructed.

Facade elements

The elements that make up the facade are perhaps the most complex, as the section is required to twist to maintain its outer edge parallel to the facade.

The full pattern

The implementation stage of this project has relied on a wide range of visualisation tools, from physical models to rapid prototypes and CAD visualisations. Without each of these studies, the exploration of the form and geometry may not have been fully resolvable.

A great deal of time was spent ensuring that geometric simplicity was maintained as well as being able to satisfy the required architectural aesthetic. The computer visualisations help constrain the geometric aspects of the members, while the physical modelling represents a tactile exploration of the potential results. This mixed-media approach allows all members of the team to participate in the creation of the stadium, and helps ensure that, as the design develops into the construction stage, the principles involved in describing the geometry can be passed logically from the designers to the constructors.

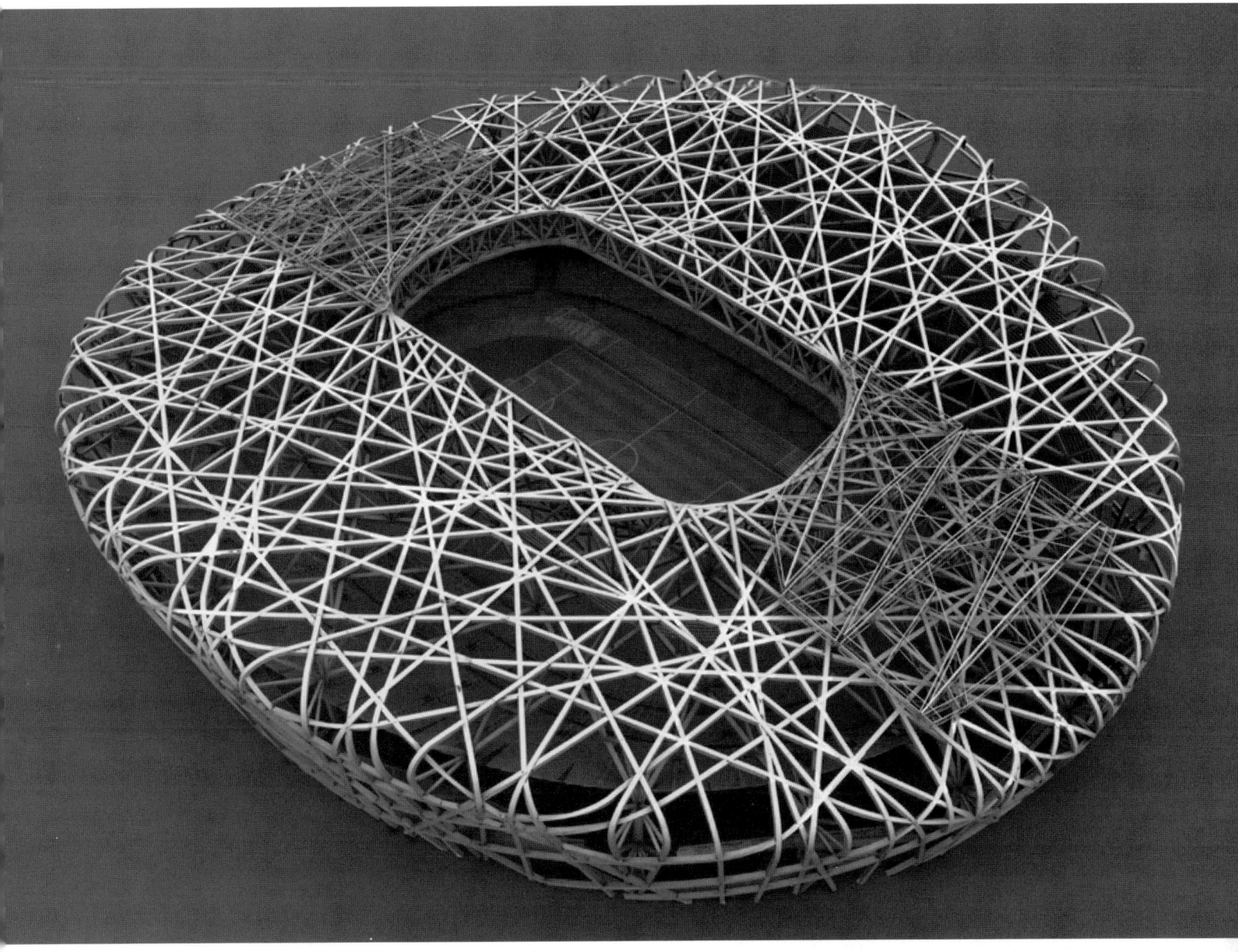

The Smart Wrap Pavilion

Kieran Timberlake Associates

The SmartWrap Pavilion for the Cooper-Hewitt National Design Museum in New York by Kieran Timberlake Associates employs new materials and digital design to create *a provocation to designers and architects*.

The combination of new materials and digital design has a transformative potential, providing building products and architecture tailored specifically to the clients' needs and site requirements. This is the essence of the architecture of mass customisation or personalised production. How can one demonstrate this physically when in essence the product is significantly ahead of current production capabilities? This was the dilemma faced by architects James Timberlake and Stephen Kieran of KieranTimberlake Associates, when asked to design a pavilion for the Cooper-Hewitt National Design Museum in the autumn of 2003. Their response is the SmartWrap Pavilion. The SmartWrap concept will deliver shelter, climate control, lighting, information display and power with a printed and layered polymer composite. The aluminium-framed pavilion is clad in a printed skin based on a combination of polyester and its derivative polyethylene terephthalate (PET), which was developed with DuPont. The pavilion was designed using a single project model, and all the aluminium extrusions of the frame were barcoded. This coding defined their structural and construction properties. This enabled an automated manufacturing bill of materials to be produced and facilitated the placement of components on site.

The fully developed SmartWrap will incorporate phase-change materials to mimic thermal mass and act as thermal regulators. Phase-change materials can store five to 14 times more heat per volume than naturally occurring latent heat stores such as water, masonry or rock. Lighting and information displays are delivered by organic light-emitting diodes (OLED), which are thin, flexible and self-emissive. The OLEDs were developed with DuPont Displays. Shelter is provided by the PET film, which is colourless and transparent. Entrapped pockets of aerogel, supplied by Aspen Aerogels, provide thermal resistance. This layered assembly achieves the thermal resistance of an insulated 400mm concrete-block cavity wall at approximately 1/100th of the weight. Noting the major power blackout recently experienced in North America, including New York, it is reassuring to note that SmartWrap also incorporates organic photovoltaic cells to generate electricity.

The museum's exhibitions curator, Matilda Mc-Quaid, stated that the pavilion 'is intended as a kind of provocation to designers and architects'. She also noted that SmartWrap should be available commercially within five to 10 years. The main pavilion is printed with a mass customisable pattern, which represents the proposed layered assembly. A 2.4m2 panel next to the pavilion demonstrates the working components of SmartWrap. Although this is currently a future product, it demonstrates a highly integrated use of nanotechnology, which can be tailored for specific site conditions and a wide range of building typologies.

Credits

Architect
KieranTimberlake Associates: James Timberlake, Stephen Kieran, Christopher Macneal, Karl Wallick, Christopher Johnstone and Richard Seltenrich
EngineerS
CVM (Philadelphia) and Buro Happold (New York)
Industrial collaborators
DuPont, Skanska (US), ILC Dover and Bosch Rexroth

Studio Work Stations and The Katsushiro Chair

Donald McKay, University of Waterloo School of Architecture

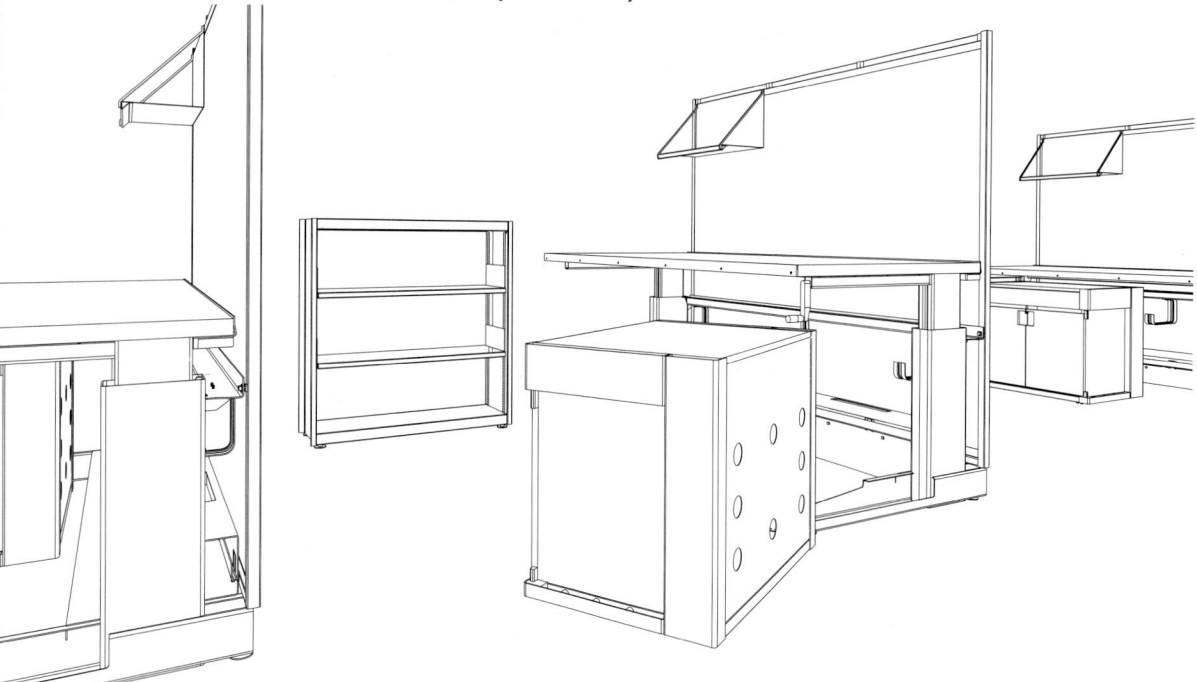

The studio workstation supports three activities: drawing, at a traditional drawing board; working at a computer – either a desktop or laptop model; making small models with light hand tools. In addition, the workstation should also operate as a study carrel and a secure storage locker for a student's effects.

In addition to a height-adjustable chair, the work stations are made up of three components, each a separate, free-standing piece of furniture:
    a table with adjustable tabletop, and a backboard;
    a rolling storage locker that can accommodate, among other things, computing equipment, and can double as a second, small table;
    a bookcase.

For each undergraduate, the school will provide a table with a backboard, and a rolling storage locker. For each graduate, the school will provide a table with a backboard, and either two storage lockers, or a rolling storage locker and a bookcase. Each faculty member is entitled to a table, with or without a backboard, a storage box, and two bookcases; faculty may choose, once, to take a credit to the value of the studio furniture for other furniture. This is in addition to any other equipment and furniture that is required in the faculty office.

The School has prepared preliminary designs for each of these components. These designs may be developed in several ways, but any redesign should meet the overall dimensions and critical internal dimensions, criteria for operations, durability, ease of maintenance; and economy. The particular configuration of the parts, the methods of manufacture, the finish, and the manner of operation are open to redevelopment.

The Table – Studio Workstations

The table consists of a tabletop, a table frame, and a backboard. It is meant to accommodate work at a computer screen, drafting at an separate drafting board that will sit on the table and can be stored under it, and a limited amount of model making. It is a student's personal desk for the term or the year, and it will take an enormous amount of wear.

- The tabletop is 36" by 72" and at least 1 1/8" thick; it adjusts in height from 28" to 42". Built of maple butcher-block (or equivalent), the top spans its frame and will not deflect under a 250 pound load. All the edges are radiused 1/8" except the leading edge, which is beveled on the underside to three quarters of an inch, and entirely rounded off. The material should provide for:

  strength and durability in the face of long wear,

  resistance to liquids,

  longevity and the capacity for renewal or restoration.

- The table frame is laser-cut, folded, welded sheet steel. The gauges will vary according to the demand on the particular component, but all of them will be heavier than conventional metal furniture components. The frame cannot fail under any amount of use, no matter how unconventional. The frame has two special demands on it:

1. it must carry power, and also (although not initially required) communications. The power requirement would integrate a four-to-six plug

power bar into the frame, along with a fifteen-foot, heavy-duty extension cord. The communications function is only projected for now.

2. it must provide the dependable means necessary to raise and lower the tabletop, either by:

   a governed electric motor,

   a hydraulic leveler,

   a counter-balanced or counter-balance and sprung pantograph,

   or a worm screw and crank.

The frame should be left as fabricated, with a clear catalytic-lacquer finish. Wear on the frame should be virtually indistinguishable from the original welding, cutting and grinding scars.

The desk may need a side panel, for privacy. This should be attached to the frame, not the top.

Provide for a strong, demountable connection from desk frame to desk frame.

- the backboard consists of a frame of laser-cut, folded, welded sheet steel, with a pinboard panel – a sandwich of two thin sheets of tentest either side of a masonite core. Mounted in place, the frame should reach to 5"3" high. The pinboard panels should be pre-painted or sealed to counter disintegration of the material. Special considerations include:

  four reinforced locations for drafting lamps,

  a folded sheet steel shelf for each backboard,

  a strong, demountable connection backboard to desk, and perhaps backboard to backboard.

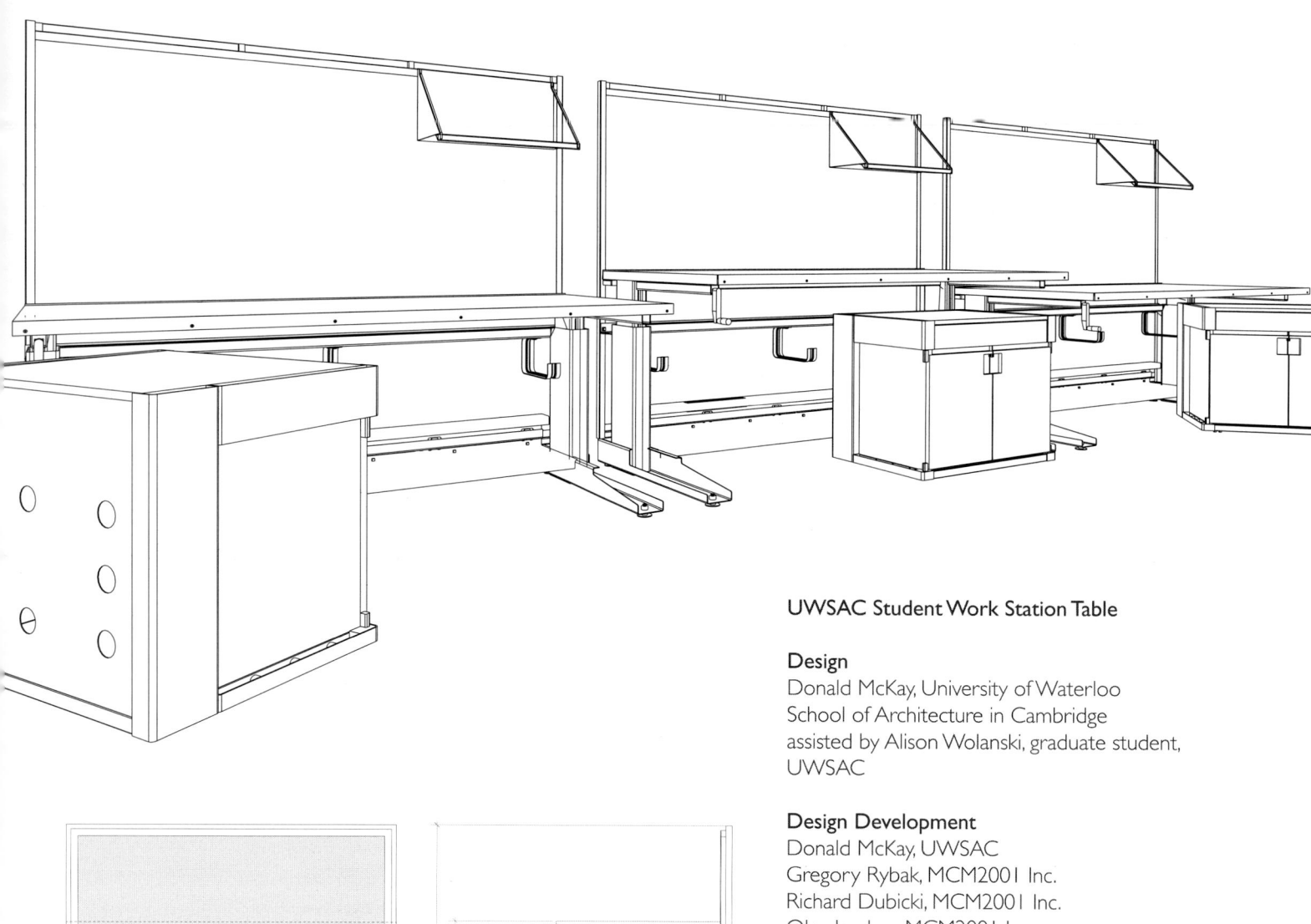

UWSAC Student Work Station Table

Design
Donald McKay, University of Waterloo
School of Architecture in Cambridge
assisted by Alison Wolanski, graduate student,
UWSAC

Design Development
Donald McKay, UWSAC
Gregory Rybak, MCM2001 Inc.
Richard Dubicki, MCM2001 Inc.
Oleg Isvekov, MCM2001 Inc.

Design Review
Building Committee, UWSAC

Prototype and Manufacturing
MCM2001 Inc.

Production Drawings
Oleg Isvekov, MCM 2001

Client
University of Waterloo School of Architecture in
Cambridge and the City of Cambridge

Perspective Views
Mark Cichy, graduate student, UWSAC

The Katsushiro Chair; The Cloud.
chair component of a three-part workstation prepared as a concept design for Teknion

Text of an email from Donald McKay to Beverly Horii, Teknion; Sunday, September 19, 1999.

Beverly:

If my notes are right, you are leaving for a holiday soon. I would love to show you what I am doing this Friday afternoon, so if we can make a date, that would be great. The following Monday would be good too, but the work will be delineated by then, and harder to backtrack. I'm happy to have you come here, or if you like, I will be equally comfortable presenting up at your office. This would be a great moment for Geoffrey's input as well, but I'm also comfortable seeing him after you've gone.

It will only be plan, section, elevations and some details in pencil for each piece -- they're nicknamed the Seven Samurai around here -- because the computer renderings, delineation and presentation will all be off in the works, but I can also explain how those things are going to work (at least roughly).

Confidentially the Seven Samurai are:

1. a mast (which takes the place of a desk) for the computer equipment, including a case for the server (which is the base and ballast for the mast), and a shroud for the mast (a sort of soft equipment case); two such masts (combined with 6. below) can make the legs for an electronic table;

2. a floating keyboard tray which attaches to the mast;

3. a floating matrix for a flat-screen monitor which attaches to the mast;

4. an adjustable screenshade-cum-lightdiffuser, which also attaches to the mast;

5. a chair-cum-recliner, for sitting and for stretching to relieve keyboard tension and to take short breaks in throughout the day;

6. a hard-and-soft chaseway, which works in a wall-mounted and a free-standing application, and also generates an incidental table;

7. a curtain system of partitions, based on the chaseway and the mastshroud.

Primary materials are:

1. aluminum castings (with a sand-cast finish, hard anodized to a dark,

   mottled gray) for bases and counterweights;

2. stamped and folded sheet steel, finished almost-black and almost-flat, like an oil-quenched steel (maybe even oil-quenched, but I'm sure that's a pipe dream);

3. bamboo (yes bamboo) set up so that it can be used in an industrial application;

4. linen-like man-made fabrics, normally used for outdoor applications -- i.e. they wear extraordinarily well;

5. "pull-trusion" glass fiber members, in an aluminum-like resin finish;

6. neoprene sheet, a bit like the stuff they make modern wet suits out of, but thinner.

7. thin-walled, oiled-cedar (or cedar ply) casework (limited, but effective);

8. plasticized ivory-white paper for the shades;

9. basically, one fastener for all adjustable parts.

The goal is to provide furniture which can make personal workstations, divide space easily, and be moved around readily without technicians, and without looking utterly systematic; which is fabricated from a limited number of parts, parts which are coordinated to be subtly intriguing in their materials and their forms and in the combination of these things, and to be resilient and simple; and which provides for a novel, compliant, comforting and comfortable personal environment for workers in a perfectly paperless office, and even in one where there is still some paper.

Client's Design Statement - short version

Designed for environments where almost all work is embodied in the computer, Meadow relieves the strain that accompanies hours at the screen and keyboard. Exercise large muscles, shift sitting position, stand, stretch; Meadow restores those activities we took for granted before keyboards, before screens, before the time and space of work became limited only by the boundary of human attention. Acknowledging that there is no single position for work over the course of a day, Meadow is designed to give us back our bodies. With the Mountain easel, an operator can move a flat screen and a keyboard to any position from crouching to standing, in a moment, without effort and without distraction. In the Cloud chair, the sitter can stretch in a full recline, relaxing back and neck muscles, adjusting posture significantly. The River viaduct makes a curtained partition of a chaseway that brings power and communications to a group of easels. If Mountain and Cloud acknowledge that frequent change is necessary for the individual's well-being, River acknowledges that change is also necessary for the group.

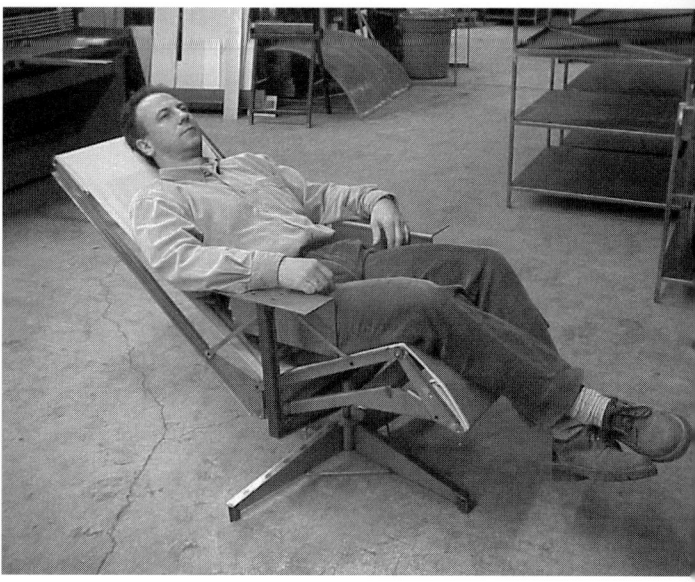

Designed in the robust language of mid-century industrial production, the Meadow suite might have found itself at home in Paris in the thirties, California in the fifties, or New York at century's end. Together, the components of the Meadow suite create a new computing environment, one where unforeseen demands on the body are acknowledged, one where the body is more at home.

Donald McKay has held a teaching position at the University of Waterloo School of Architecture since1979. He is principal of a design and planning studio in Toronto, first established in 1989; Donald McKay – studio works in furniture and equipment design, architectural and urban design. Beginning in 1993, he began research on American culture and public life. He is now completing volume one of this work – Abysmal Crossing – in which he describes spaces in the mythic experience of western America. In November 2004, mural-scaled prints of his photographs for this book will go on exhibit in the David Azrieali Gallery at Carleton University.

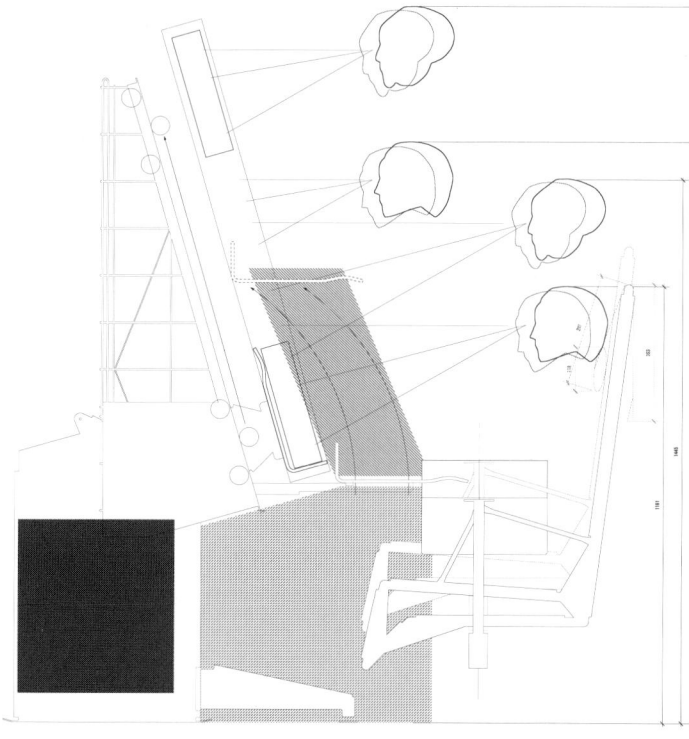

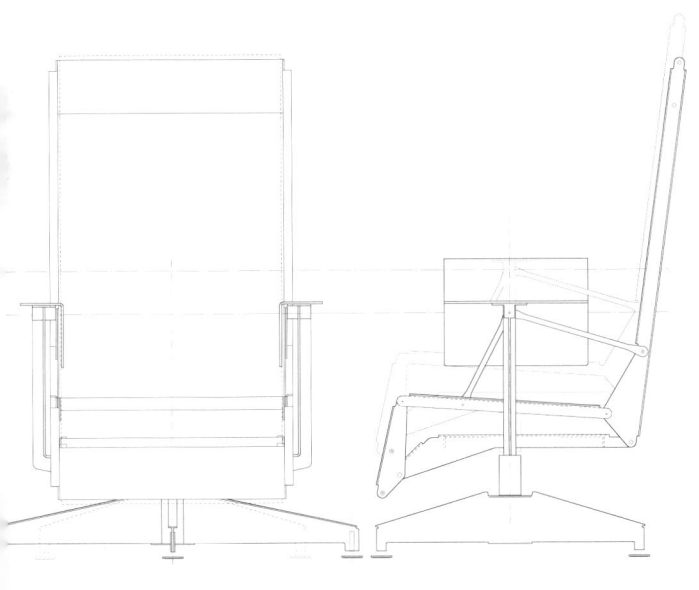

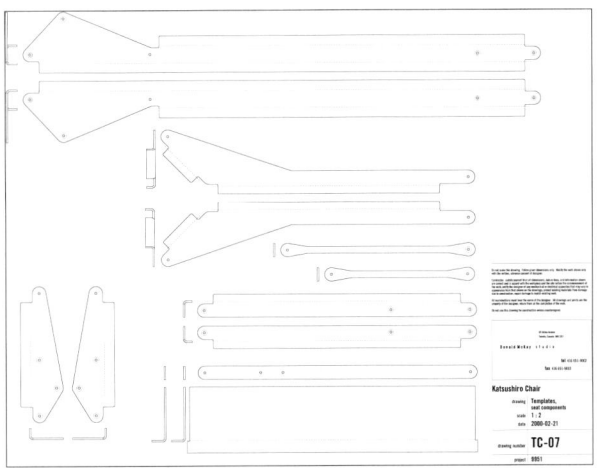

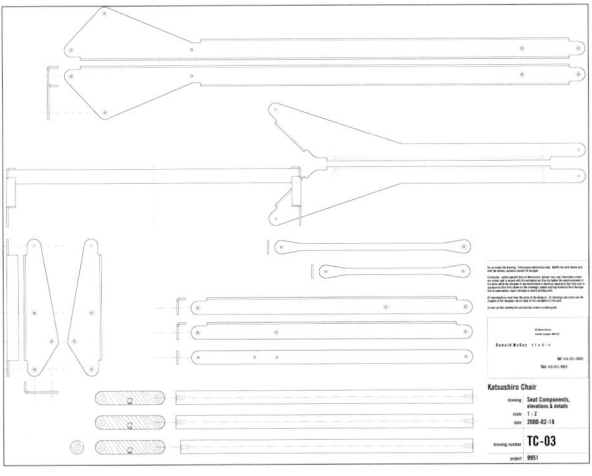

Postagriculture

Achim Menges

Behind the current benign image of agriculture lies a complex organisation of production processes that is powered by an enormous energy consumption, slow to respond to change and dangerously inflexible. In the Netherlands these problems are exacerbated by two factors: firstly, there is higher agricultural productivity per land area than in any other country in Europe; and secondly it is the most densely populated nation in the European Union. This places severe pressure on the Dutch landscape; a necessity for increasingly high agricultural productivity in conjunction with very dense population.

The ambitions of the 'Postagriculture' project, by architect Achim Menges, stem from the recognition of the critical importance of environmentally and socially sustainable food production. The aim is to develop an inclusive and responsive strategy that will enable a mode of agriculture that is highly integrated, mutable and a vital urban programme.

Interrelated systemic, regional and national investigations of agricultural production and the density of population make it obvious that an alternative strategy has to promote the intensification of already existing agro-production and at the same time provide for public recreation in close proximity to urban areas. Consequently the central aim of the strategy is a local hybridisation of intensified agro-production with public recreation. Therefore an organisational model must be developed that is capable of negotiating different system requirements in an adaptive manner.

Condition-basedorganisation

Through a digital technique based on parametric information of light and climatic requirements of various agricultural and recreational systems, an organisational model was derived that indicates specific intrasystemic condition profiles and differential inter-systemic relations. Within this new model, system specificities emerge as patterns of gradient thresholds and potentially shared conditions. As it is dynamic, the model is an operative tool for evolving programme deployment, organised by differential intensities of light, temperature and smell. Thereby inter-systemic relations are mutable and occupations are time-based. This allows the organisation to respond to differential demand for agricultural products and the changing needs for recreation areas.

The abstract machine of this model operates on a topological base of inter-systemic relations and connections, and is essentially scale-less. It can be employed to test internal developments and changes, and to absorb and reconfigure new information and external expertise at any stage of the project. A continuous feedback loop is established, enabling an intricate dance between analytical and generative modes.

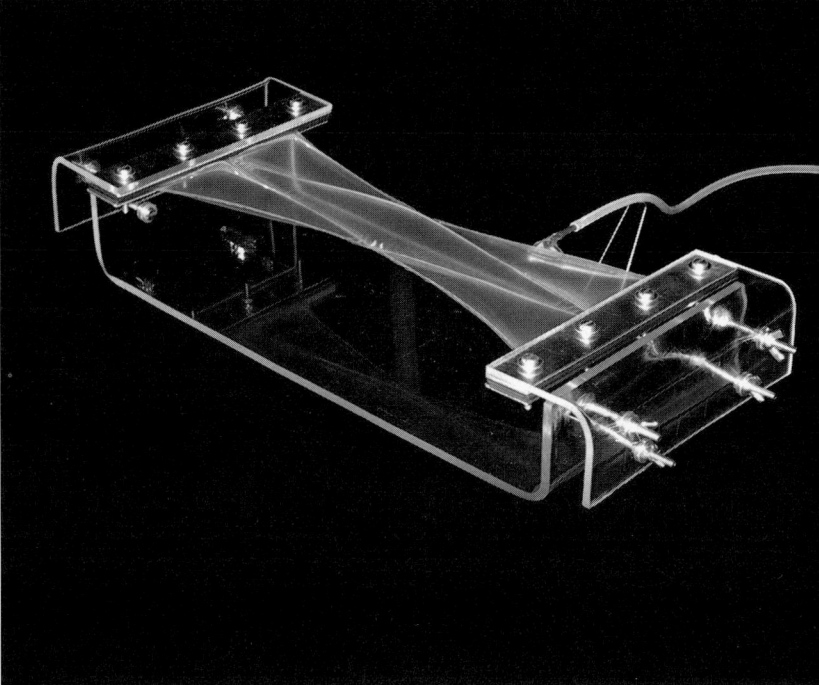

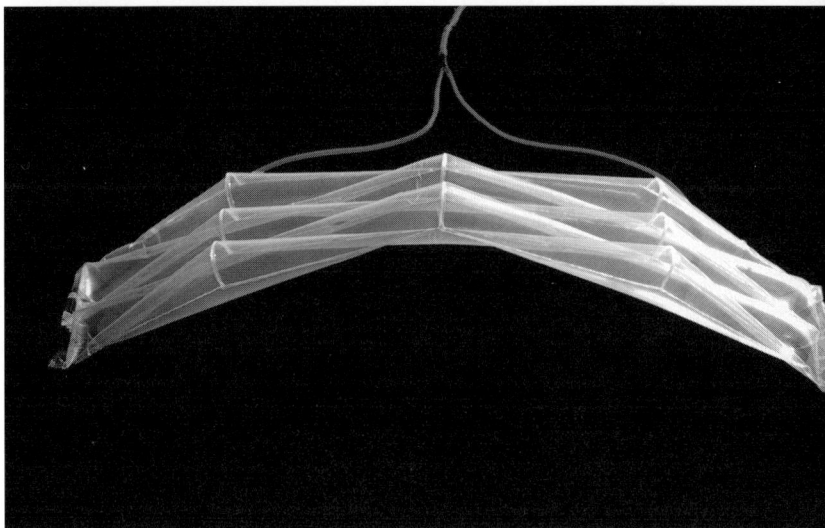

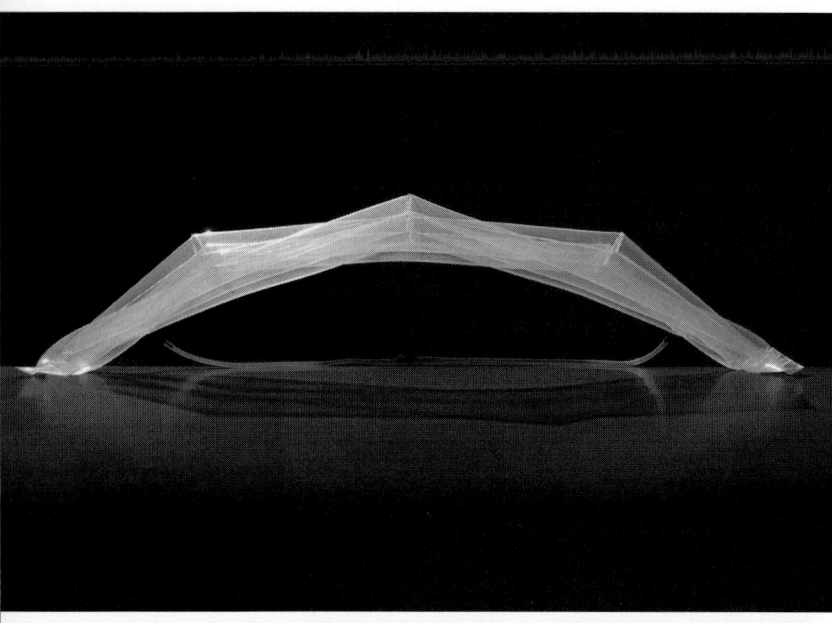

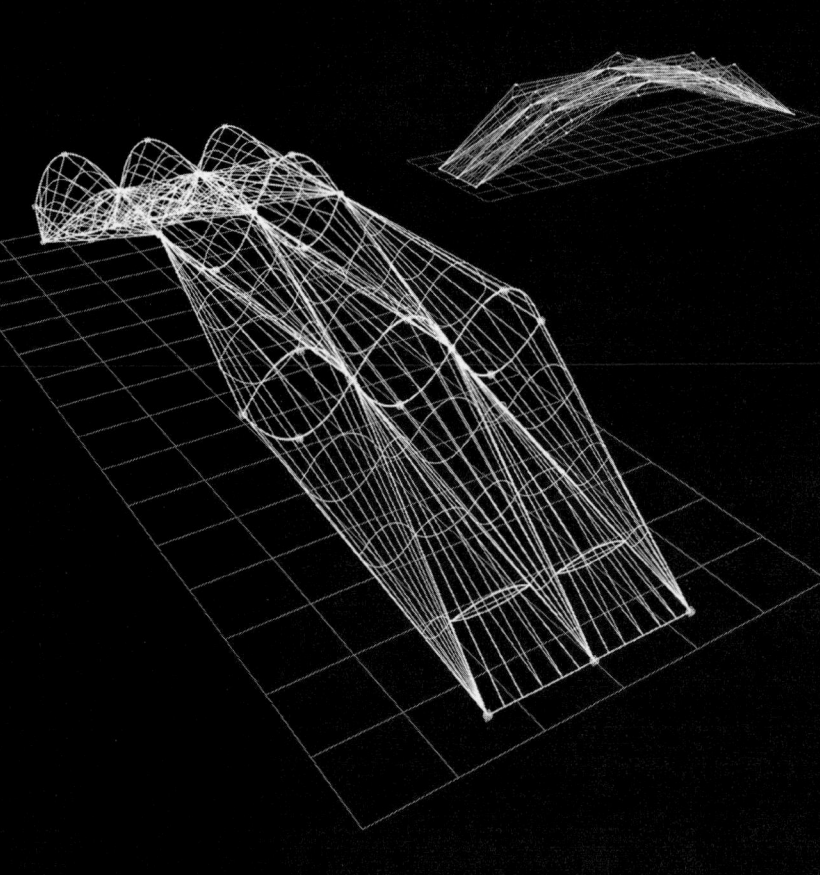

Deep structure

A performative environment necessitates thinking about structure as a condition that generates and differentiates. Rather than a static object, it is helpful to think of structure as a process of structural and material operations. Manipulations within the organisational logic and constraints of the structural system are intrinsically connected to the modulation of microclimatic conditions. In fact, the structural articulation enables the organisational model of differential light and climatic intensities to become operative. Therefore the development and assessment of the structure is not simply limited to its load-bearing capacity but to a whole set of other performance criteria.

Pneumatic structures were explored, as they have great potential for differentiation, and this was achieved by exploiting their non-linear characteristics and differential states of stability. The working methodology was based on feedback between different modalities – a digital definition of the boundary points and the related cut patterns produced in engineering software, physical modelling, digital form-finding and digital structural analysis. This process evolved a specific pneumatic component, which was further developed towards a self-supporting pneumatic surface prototype. Parametric changes of variables – such as the orientation, distribution, density and differential reinforcement of the seams, the depth and the internal pressure of the 'pneus', and the type and treatment of the surface material – were employed to manipulate the proliferation of the surface component in relation to its structural performance and its capacity to modulate light and climatic conditions. The resulting deep structure is a manifold of local geometries within a global system.

Performative environment

The articulation of the project takes a multitude of site-specific parameters and uses them as generative forces. The site at the seafront in Westland has specific light and climatic conditions, existing local agriculture, recreational networks and infrastructure. Together, these disparate parameters inflect the distribution of macro environments and macrosystemic connectivities. This macro environmental distribution, combined with the differentiated structural system, evolves a highly articulated post-agricultural landscape.

In the microsystemic layout the livestock and solar gain are the main heat and light sources, conserved and modulated by the thermal mass of fish basins. A digital simulation run of one year, with particular focus on temperature scan and light analysis, indicates how the structure responds to locally specific load-bearing requirements and evolves microclimatic differentiation and thus enables the organisational model to become operative. More importantly, the analysis of the structural and environmental performance also indicates the gradient transition from the hard-modulated micro environments to a negotiable field of soft-modulated areas. These areas, which emerge between the regulated material and surface manipulations, enable a robustness of distributed open systems over a broad range of conditions that can trigger and accommodate a manifold of programmatic mutations. This indicates that the project follows a twofold strategy: one passive and one active. Firstly, the negotiable field of differentiated micro environments passively provides for anticipated criteria of change. Secondly, active key structural elements provide adaptation for divergent criteria. The consequent mutations of relations between systems will then feed back alterations to the topological organisational model.

In conclusion, the project demonstrates how an intensified localisation of agriculture could save future resources by including urban programmes in a production landscape. It aims to clarify the critical agenda of agricultural production, presenting an argument for a paradigmatic shift away from the current practice of concealing industrialised food production. It is self-evident that this requires a structure and an organisation that is responsive, and capable of accepting programmatic, spatial and environmental changes to accommodate fluctuating needs.

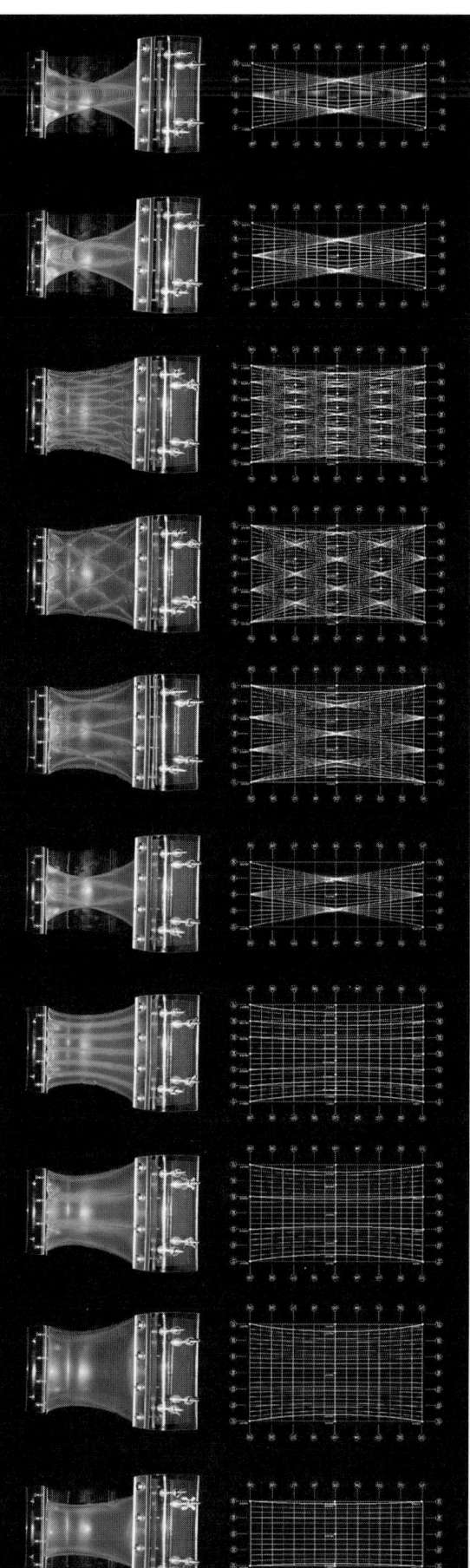

The Nasher Sculpture Center, Dallas

Renzo Piano Building Workshop

The project consists of a two acre sculpture garden and a gallery building in downtown Dallas. The brief was to create a home for one of the most significant collections of 20th Century Sculpture in the world. The two-storey gallery building contains an auditorium, restaurant and workshop facilities along with 1000 m2 of gallery space.

One of the driving concepts for the building was to provide as much natural daylight to the principal internal galleries as possible to provide optimum viewing conditions for the artwork. The desire to maximise the levels of daylight is constrained by the impact this would have on the building's cooling load and also by the sensitivity of many of the items of artwork to direct sunlight.

The concept developed with the architect was to have a high-performance glass roof over the galleries with a carpet of external shading elements above filtering the sunlight and providing solar protection.

Energy Efficiency

The roof provides a high level of transparency without an energy penalty by achieving a shading factor of 95%. This still allows the upper floors to rely mainly on sunlight for an energy efficient lighting scheme. A floor displacement system in the upper levels takes advantage of the opportunities for free cooling on mild days.

Multidisciplinary design

The glass roof means that high level services have to be minimised. Lighting tracks are incorporated into the roof support structure to provide a sympathetic but flexible lighting scheme that works with the daylighting. Building services were integrated into the structure in order to minimise their impact on the architecture as illustrated below.

New Technologies

Arup went back to first principles to develop an innovative shading solution for a building where architecture and aesthetics are valued. The development of the design began by calculating critical solar angles to determines the exact curvature of the shading shell opening that would prevent any direct sunlight penetrating the shade at any time of year. The decision was taken to work with three-dimensional computer models for the development of the form allowing the integration of the complex solar data with the geometric models. Throughout the development all information was communicated through the exchange of 3-D models; no paper drawings were produced.

3-D modelling software produced sophisticated renderings of the shading but this wasn't enough to fully explore the geometry so Arup invested in in-house rapid prototyping facilities. Full-scale, accurate physical prototypes could now be produced directly from the 3-D computer models within hours. This was a major benefit to the communication of ideas between the design team and client; even subtle changes to the geometry could be studied. The models also allowed the production of aluminium prototypes through a process of sand-casting.

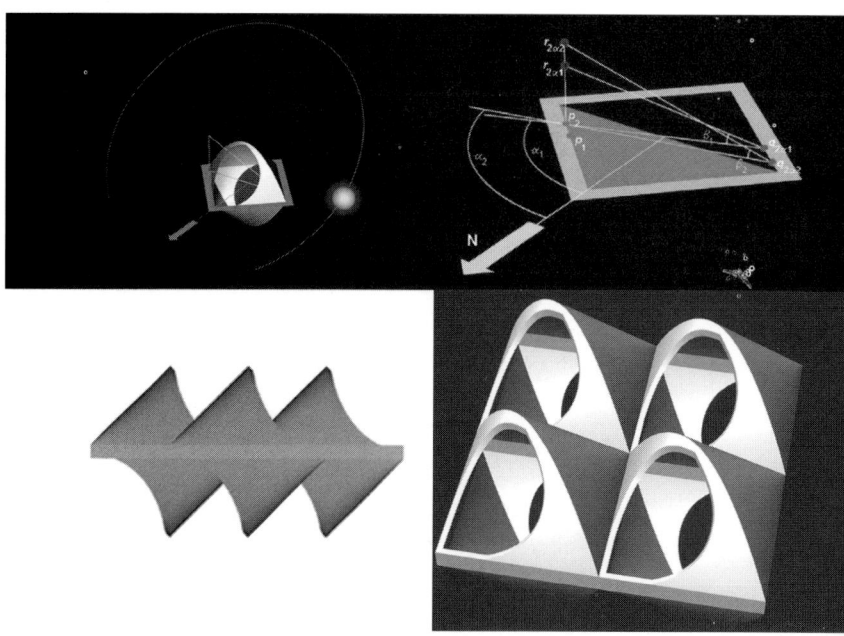

The innovative use of 3-D computer modelling and rapid prototyping enabled the development of this unique shading product optimised for the site. It enabled good communication with the client and design team and dramatically reduced the time from concept to manufacture.

The gallery opened in October 2003.

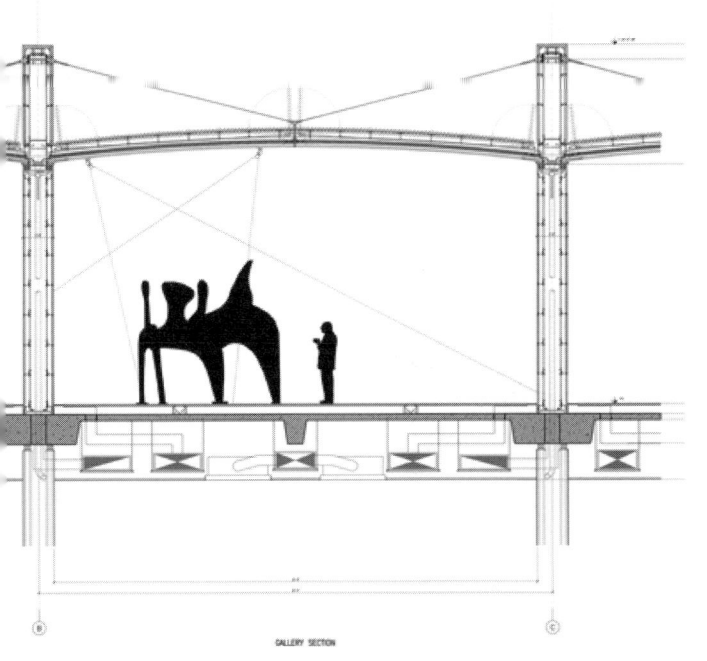

GALLERY SECTION

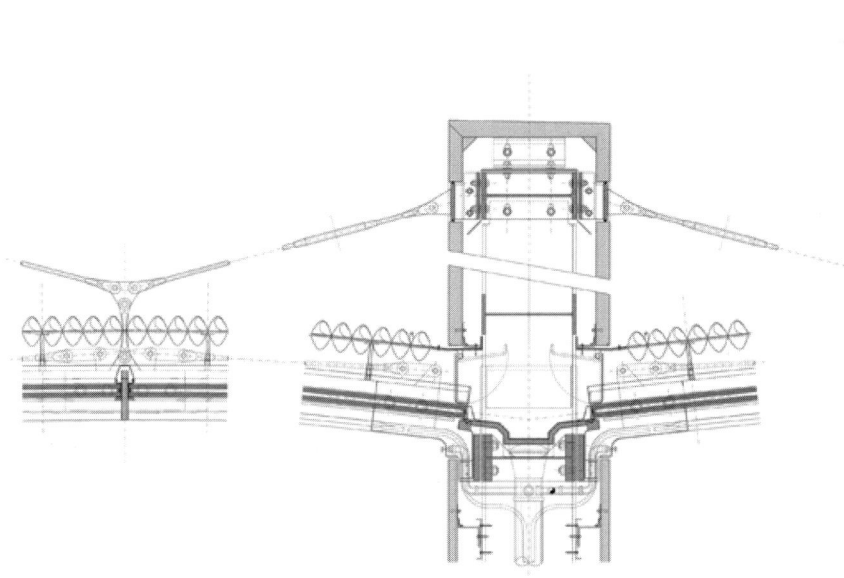

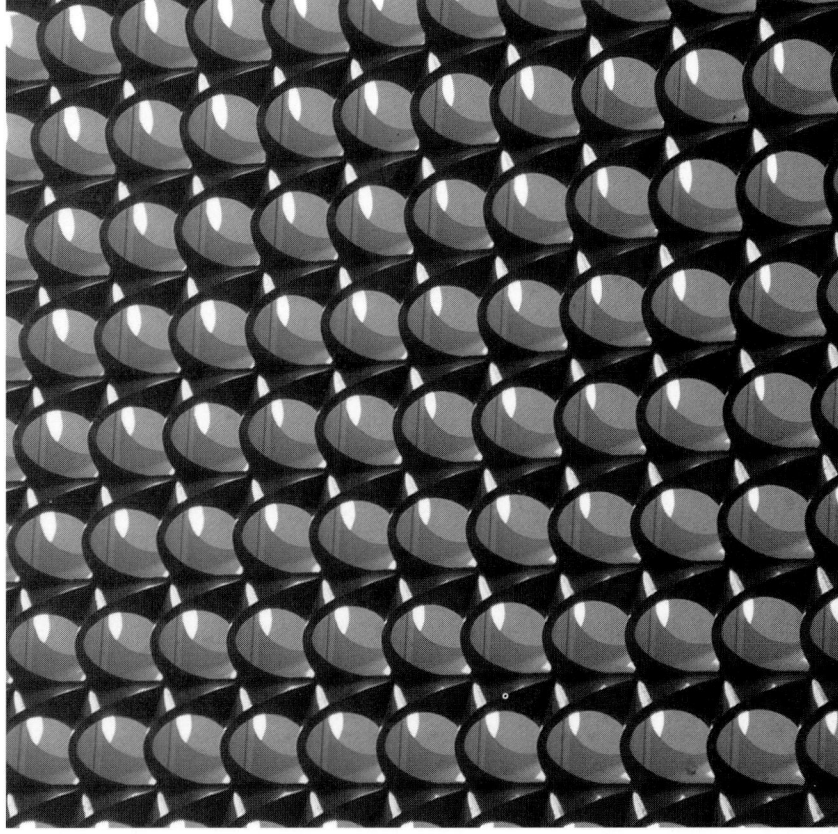

Antwerp Law Courts

Richard Rogers Partnership

Richard Rogers Partnership's design for the new law courts in Antwerp elegantly demonstrates the progression from digital representation to digital fabrication including large-scale off-site manufacture

The new law courts complex for the Flemish city of Antwerp is one of Richard Rogers Partnership's (RRP) major public buildings of the early 21st century. Like many projects by the practice, it reflects a vision of the city as a humane and inhabitable democratic place and represents a commitment to the regeneration of urban life.

The site for the law courts is at the Bolivarplaats, on the southern edge of Antwerp's central area, where the urban fabric is broken by a massive motorway interchange, cutting off the boulevard that leads into the city. The new building is one of the catalysts for RRP's long-term masterplan of 'the new south' of the city, currently in progress.

The courts building, designed in conjunction with Belgian co-architect VK Group and Arup, is conceived both as a gateway to the city and as providing a link across the motorway between the city centre and the Schelde river. It houses eight distinct civil and criminal courts and includes 36 courtrooms plus offices, chambers for judges and lawyers, library and dining room, and a great public hall – the 'Salle des Pas Perdus' – linking six radiating wings of office accommodation. This space is capped by a striking, crystalline roof structure. The courtrooms are placed on top of the office wings and are crowned with hyperbolic paraboloid roof forms.

A low-energy services strategy is fundamental to this project – natural light is used to optimum effect, reducing reliance on artificial lighting. External solar shading to the office facades controls the solar gain. Opening windows provide natural ventilation to the offices, supplemented by low-velocity ventilation for the hearing rooms. Rainwater is recycled.

The building, straddling a major highway, looks out to a large area of parkland and is designed to create 'fingers' of landscaped parkland, which extend right into the heart of the building. The landscape is configured and planted to shield the building from the noise and pollution of the motorway.

Courtroom roofs

Arup modelled the roofs in ACAD 3D, importing the centreline geometry and sections sizes from OASIS GSA, which is the Arup in-house structural analysis package. Each courtroom roof is composed of four geometric hyperbolic paraboloid (HP) forms. In simple terms, the HP is a double-curved surface: on plan it consists of a simple rectangular grid with the corners pulled up (or down) to create a double curve. On each courtroom roof, two HP forms are 'pulled' higher and cantilevered over the two lower ones, creating a gap in-between, which in turn is glazed to maximise the natural light coming into the courtrooms. These rooflights face northwards, with the overhangs providing solar shading against the high-altitude sun. The four roof sections are seen as individual, identifiable components and are further separated by strip rooflights between each of the two higher and lower roof elements.

The initial design of the HP forms consisted of a perimeter steel tubular frame containing short sections of prefabricated laminated timber beams connected at the nodes to form the double-curved

grid. This grid of beams in turn was covered with LVL kerto plywood to form the outer structural skin of the roof.

Following detailed discussions with the timber contractor, Merk, a new solution emerged, consisting of grid beams laminated in full lengths with each layer. This arrangement was built up progressively by screwing together long lengths of timber and connecting these to the perimeter tubular steelwork. The new simplified solution obviated the necessity for the original timber-to-timber connections, allowing the long timber-strip construction to follow accurately the geometry of the HP form. The original LVL ketro plywood shell was replaced by similarly long lengths of timber forming the outer structural skin of the roof.

The whole of the roof construction was assembled in a large shipyard a few kilometres upriver from the site. The work areas were set out as a production line to deal with welding, painting, timber gridshell assembly and roofing. Given the number of roofs that had to be assembled (32), working under cover ensured quality control of the end product.

Several contractors worked together. Partially prepared materials were delivered to the shipyard and assembled in an efficient manner – the first complete roof emerged in under four weeks. Thereafter, one complete roof (four sections) was being delivered to site every week.

The prepared tubular steel members for each of the HP forms were transported from the steel contractor and then welded together to form the HP's perimeter frame. These frames were then moved into a giant spray booth, where they were prepared and painted.

The frames were then moved to the next work area, where they were set on to a timber jig reflecting the double-curved surface. The timber contractor assembled the grid beams, which were built up layer by layer on the jig, making connections to the steel perimeter frame. After the required number of layers had been built up, the shell of the structure, consisting of similar strips of timber running at 45 to the grid beams, was assembled to form a continuous top surface.

The roof sections were then moved to the next work area, where the roofing contractor installed the vapour barriers, insulation and the stainless-steel, fully welded, standing-seam roofs. The cantilevers and all edges not requiring site interface were completed. Roof edges requiring later site interface were prepared and left ready to 'zip up' on site.

The complete sections of the roofs were transported along the river by barge, completing their journey on the back of a wide-loader across the fields, finally arriving on site where they were craned into position.

Solar-shading brackets

Office accommodation is organised into six radiating wings on three levels. Offices are naturally ventilated, with external solar shading. This shading consists of 2.8m-long fritted glass louvres fixed to a carrier system, with access walkways cantilevered from the facade line at 1.4m centres. The brackets, which hold the glass at the predefined angle and pitch, were initially designed to be part casting and part profiled cut plate. The louvres were intended to be 'held' at the ends by the brackets, obviating the need for holes in the glass.

The cladding contractor, Permasteelisa, proposed to cast the whole bracket in stainless steel for strength and corrosion purposes, which required RRP to revise the design to streamline it for the casting process. The design was developed from the initial sketches, through to CAD drawings; a full-size rapid prototype was then made and subsequently adjusted and refined by RRP. The casting of the support brackets was eventually outsourced to a foundry in mainland China.

Design Team
RVA Joint Venture
(Richard Rogers Partnerships/ VK Group/Arup)

Client
Regie der Gebouwen

Promoter
Interbuild

Cladding Contractor
Permasteelisa; thv Gevelcombinatie Antwerpen

Steelwork Contractor
Lemants

Timber Contractor
Merk

Roofing Contractor
ME Construct

Kingsdale School

De Rijke Marsh Morgan Architects

The design team for the refurbishment of Kingsdale School used dynamic thermal-analysis software to explore the use of ventilation and light in the creation of a 3,200m2 variable skin roof.

The Kingsdale School project, a £10 million Architecture Foundation/Schoolworks and DfES-backed (Department for Education and Skills) initiative to demonstrate whether quality architecture improves school standards, is well advanced on site in Dulwich, south-east London. The roof allows for various uses – including circulation, dining, assembly, auditorium and library – within a spectacular 3,200m2 indoor/outdoor space. De Rijke Marsh Morgan Architects (dRMM), working with Mike Hadi Associates, Vector Special Projects and Fulcrum Consulting, designed the 40 x 80m ETFE Texlon variable skin roof – the first in the UK and the largest in the world. A technical tour de force, it represents Buckminster Fuller's ambition of a climatic envelope realised in a material he could only dream of.

The opportunities of the variable skin option, including integrated passive/through ventilation and light levels, were explored using dynamic thermal-analysis software.

Ventilation is via doors and motorised dampers at ground-floor level, and by ETFE flaps in the clerestory. Summertime overheating within the surrounding classrooms will be reduced considerably thanks to the shading provided by the new roof, and in winter fresh-air ventilation will be increased thanks to the manually openable windows that face into the sheltered courtyard space. The insulating buffer zone of the 'climatic envelope' generates a subsequent net energy saving to the existing building.

dRMM designed the distinctive solar shading pattern on Op Art principles, developing a dynamic three-dimensional solar control principle, offering extraordinary qualities of light; the brief called for 55 per cent solar shading generally and 95 per cent during the warmer season where temperatures were analysed as peaking around 35 C.

The amount of air to the two cells within the cushions can be altered either automatically by sensors or manually by teachers. By splitting the 23 cushions into three separate zones, another layer of flexibility was achieved, allowing for incremental alteration of the daylight within the space – say for a theatrical performance. The main roof can alter from light to dark within 30 to 45 minutes.

A span to weight ratio of approximately 30kg/m2 made it possible to superimpose the structure straight on to the existing flat roof with no foundation requirements.

Architect
de Rijke Marsh Morgan Architects

Structural Engineer
Michael Hadi Associates

Service Engineer
Fulcrum

Main Contractor
Galliford Try

Quantity Surveyor
Appleyard & Trew

Steelwork Subcontractor
SH Structures

Lattice Archipelogics

SERVO with Smart Studio/Interactive Institute

An installation by Servo in collaboration with Smart Studio/ Interactive Institute uses spatial software to translate movement patterns into lighting formations.

Lattice Archipelogics, by Servo with Smart Studio/Interactive Institute, integrates new technologies and material systems in an architectural installation for the exhibition 'Latent Utopias' at the Landesmuseum Joanneum in Graz. The intention is not so much the static display of a material object, but the generation of new material through real-time interaction between the object and its users.

Integrating both physical and virtual infrastructures, Lattice Archipelogics functions less as a site for passive consumption and more as a dynamic spatial instrument of active production. Using the museum hallway as a corridor of circulatory exchange, it absorbs, processes, and ultimately performs the social and political interactions of visitors within the gallery itself. A suspended cellular archipelago of fabricated elements embedded with intelligent technologies – motion sensors, LED lighting, and speaker technologies – responds to, as well as influences, the movement patterns of its users. These movement patterns are then interpreted by software designed specifically to translate that information into a series of lighting formations, eventually distributing these formations throughout the lattice structure via an array of LED lighting elements. It is to this extent that the installation effectively indexes and materialises its users' activities by becoming an instrument of their individual and collective potential. The end result is as much the existing material system itself as it is the virtual inflection and transformation of that system by its users.

The spatial hardware of Lattice Archipelogics is a thick atmosphere of 102 vacuum-cast polyurethane cells. Through the numerical input of a 3D model, an acrylic STL rapid prototype provided a positive form for a negative soft silicon cast. The softness of the silicon cast, the possibility of peeling it off the positive form, made it recyclable for the fabrication of several vacuum-cast polyurethane cells. In Lattice Archipelogics the cells interlock with a twin cell in the formation of the larger system. The cells are later staggered vertically in a pattern of self-counterbalance, and double as conduit for the distribution of various cables and wires specific to its structural suspension and programming systems.

The spatial software of Lattice Archipelogics is an intricate web of interactivity. Motion sensors track the movement of visitors through the installation, effectively allowing the system to observe or listen to the behaviour of its occupants. That information is then processed by software designed to translate these patterns into new ones. When left in its inert state, the algorithmic interface will sample from the catalogue of stored movement patterns to perform them iteratively. These newly generated patterns are then distributed through the system in the form of light, altering the spatial conditions of the installation.

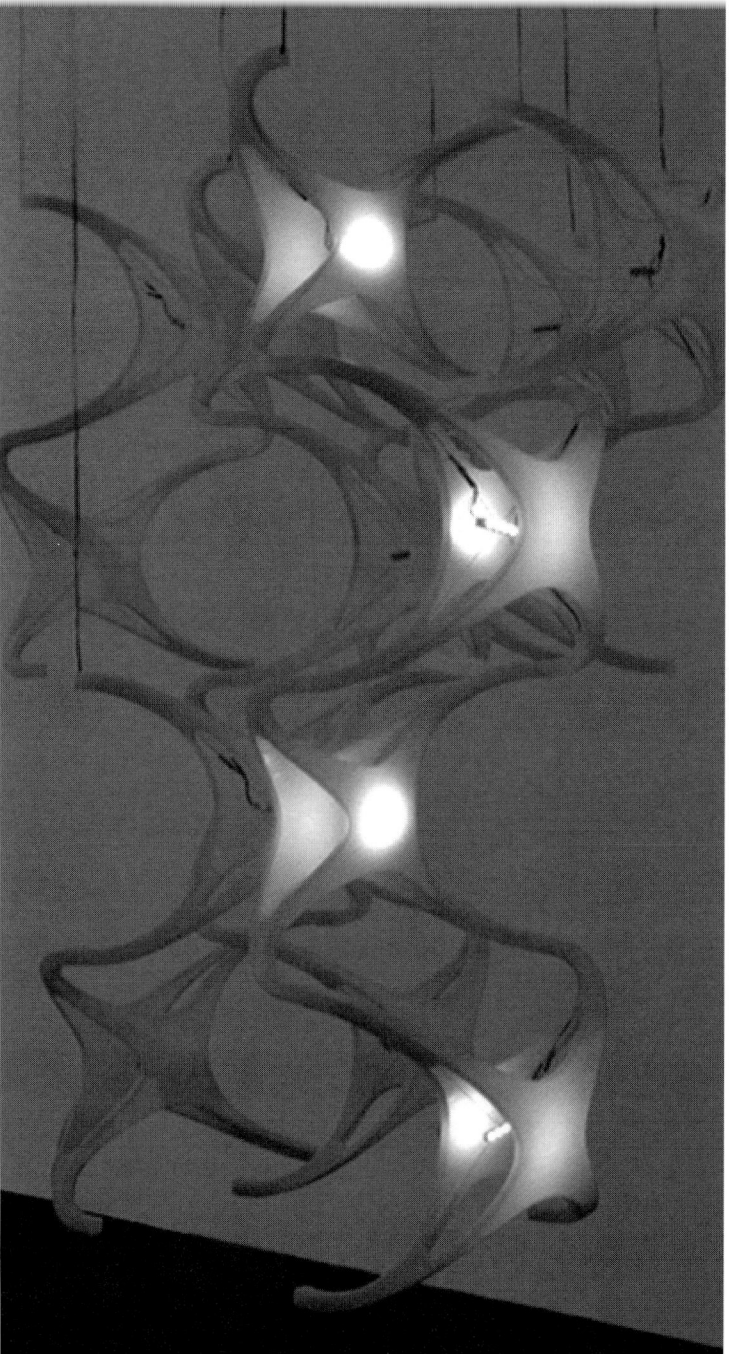

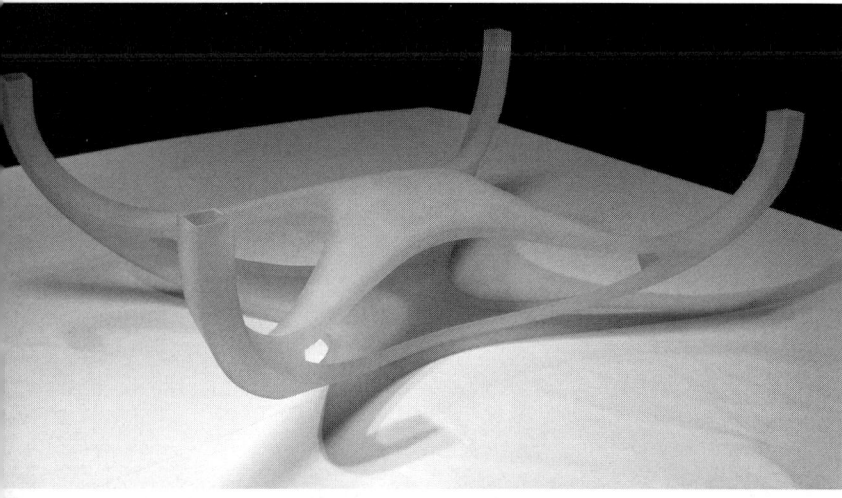

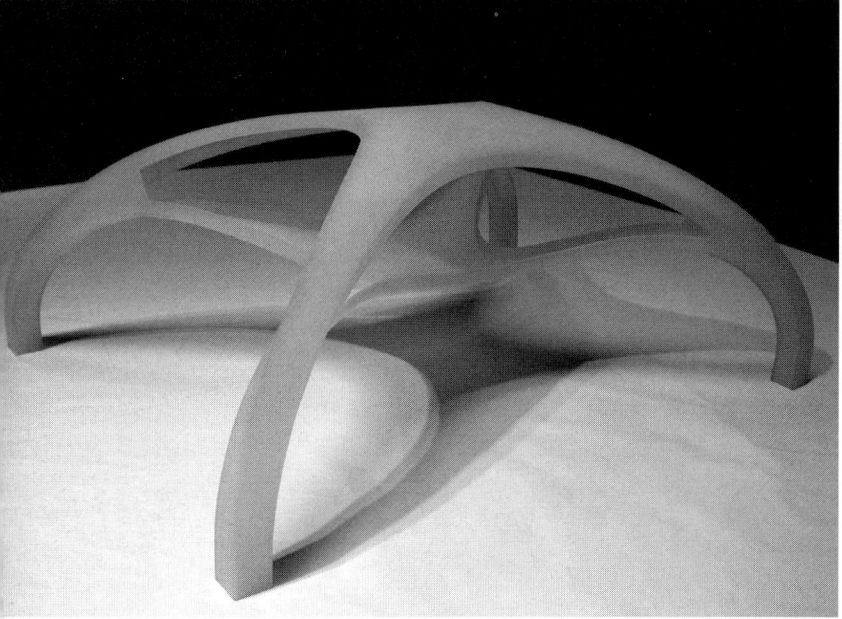

Principals
David Erdman
Marcelyn Gow
Ulrika Karlsson
Chris Perry

Design team
Servo
Daniel Norell
Clare Olsen
Jonas Runberger

in collaboration with Smart Studio/Interactive Institute – design team
Ingvar Sjöberg
Tobi Schneidler
Fredrik Petersson
Olof Bendt
Magnus Jonsson
Pablo Miranda
designers of the responsive field of lattice archipelogics. Vacuum casting and fabrication consulting provided by CARAN

Thanks to
IASPIS (International Artists' Studio Program me in Sweden), KTH-Royal Institute of Technology, SSARK medialab, White arkitekter, and CARAN

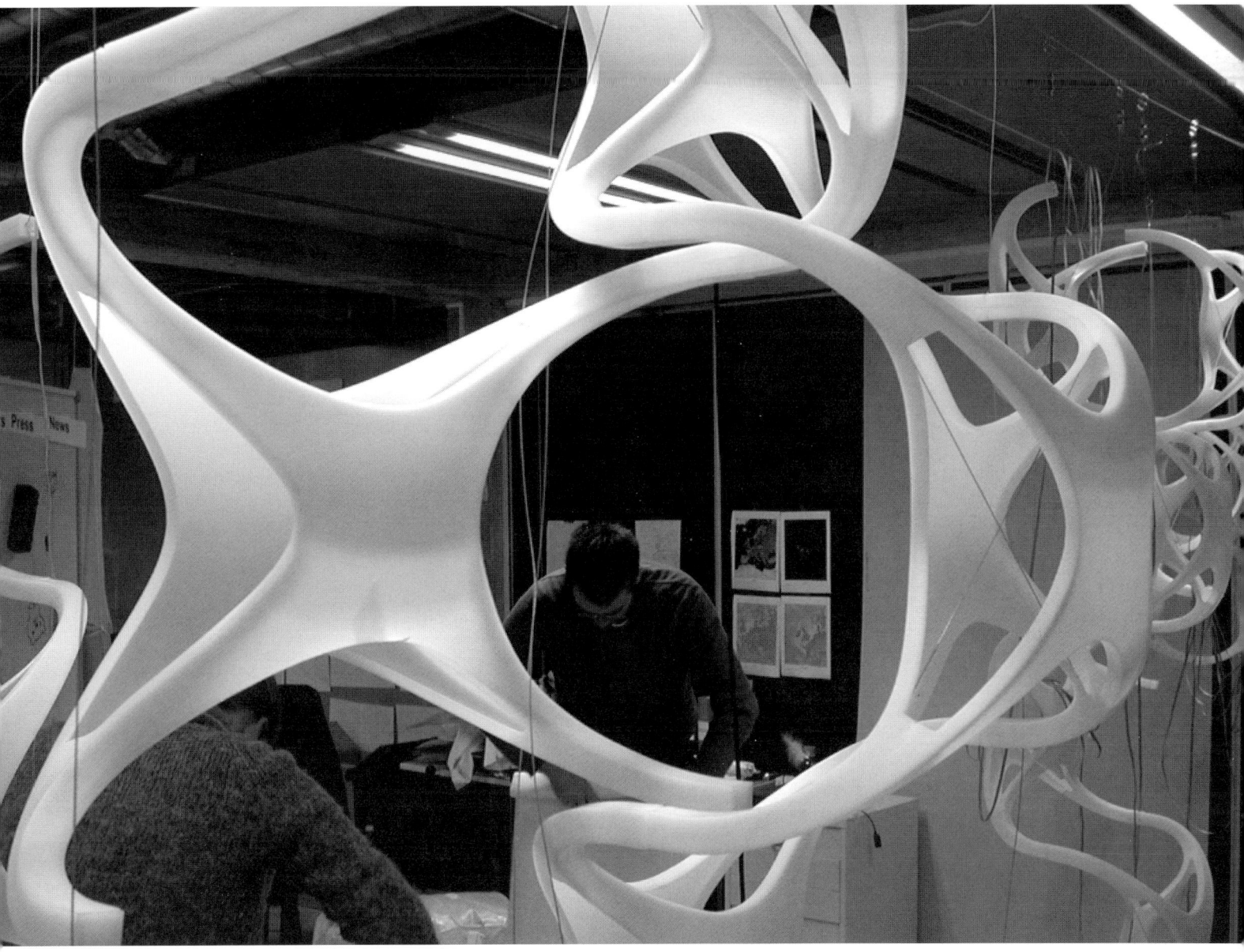

Shorting the Automation Circuit

Bob Sheil with Phil Ayres snd Nick Callicott

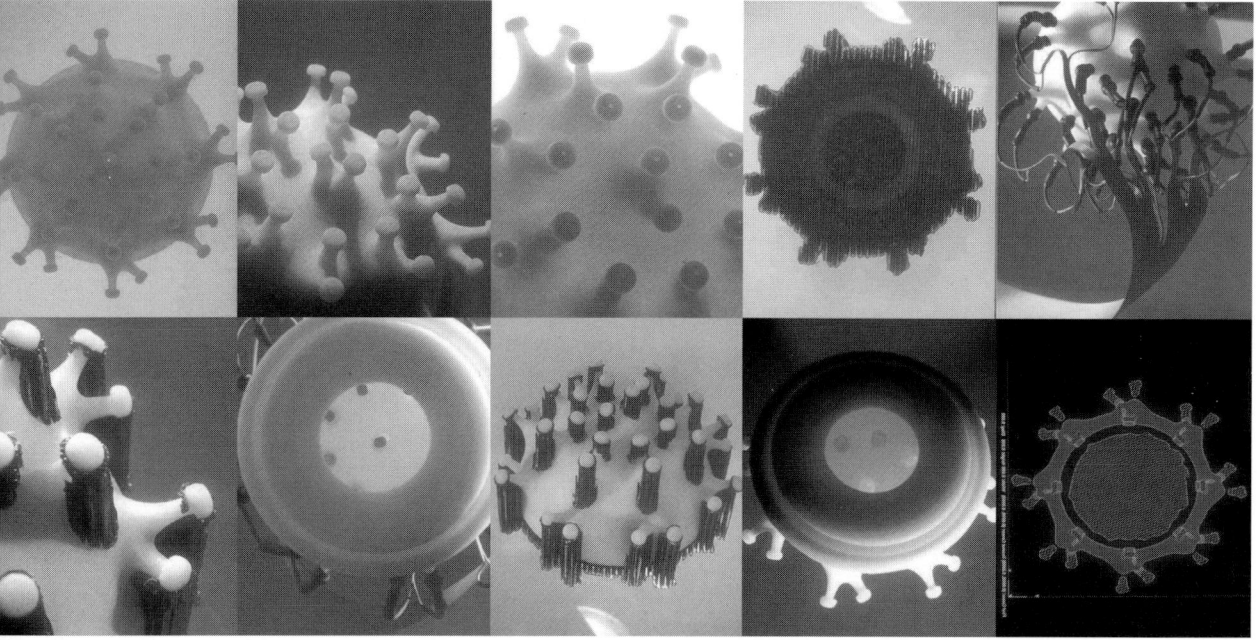

The effects of computer-aided manufacturing have often been defined simply as a hastening of the path to production, but in this site-based installation was made to re-configure that path itself, placing authors, object and environment within the 'automation circuit'.

The work exist both as object, system and event, the overlap of which creates an opportunity to develop extensions to existing design practices through a direct relationship with the manufacturing medium. It also seeks the revision of our existing understanding of production by constructing an interactive relation between site, audience and representation that is activated by the manufacturing process itself.

Process and technology

The project is sited within a redundant observatory at University College London, whose existing mechanical roof provides an articulated threshold to the surrounding public quadrangle. It is here that our object resides, reviving a tradition of observation and collection long past.

The object itself embodies a series of requirements and influences, which are functional, sculptural and contemplative. Housed within its hollow interior is an array of light dependent sensors, each of which are selectively activated by the light refracting through an acrylic conduit whose angular position is governed by prevailing air currents. The internal cavity of the object is structured by the sensor array geometry; whilst its outer surface topography reveals a further redesign develop additional sculptural properties of that geometry by exploiting the capabilities of the solid freeform fabrication process. The final object exhibits many intricate internal features, and would be impossible to fabricate as a single component using conventional machining or handcraft. Its virtual description was constructed as a solid model using a CAD package, and converted into an .stl file prior to manufacture.

In use, the object's sensors record an informal 'thumbprint' of the prevailing physical conditions, one that is also inclusive to the more dynamic presence of passing spectators. The sensor signals are digitized using a STAMPII control processor interfaced to a PC, and a terminal program then collates and tabulates the data into a text script suitable for import back into a three-dimensional CAD application.

The transference of this environmental data into the modeling program suggests a number of subsequent avenues of research. If required it could be directly converted into form, providing a legible means of representation for some environmental analysis applications. In this instance, however, the intention is more generative and seeks to integrate the information with subsequent design proposals for the site.

By combining the data with existing authored representations we can experiment with the 'weathering' properties of the data upon previously modeled form, adding or subtracting material in the search for specificity.

The approach does not constrain or prescribe our creativity, whose outcomes remain evident both in physical form and within the algorithms and architecture of the system itself, yet the underlying desire is to reconcile this authorship with the social, environmental and technological factors that comprise the reality of the site, and its means of production.

Fabrication of the fused deposition model was carried out at the Department of Medical Physics, University College London with the assistance of Dr. Robin Richards.

FDM is a registered trademark of Stratasys products.

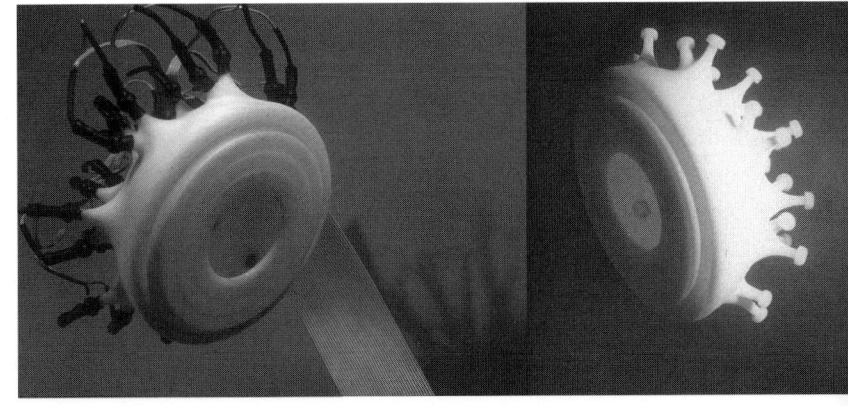

Nested House

Urban Future Organization

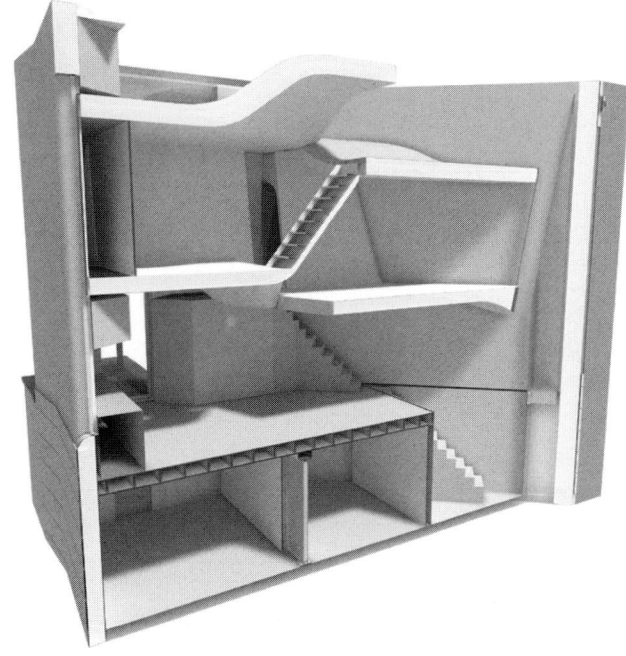

A small residential project in Hackney uses one digital three-dimensional model as the single source of all project information.

The site is an existing 45m2 maisonette terrace conversion distributed over three floors in the heart of the De Beauvoir Conservation Area. Due to the conservation area restrictions, the designers deliberately decided to concentrate on the reorganisation of the interior space and its relation to a split-level garden on the roof, offering intimacy and privacy as well as unobstructed views to the City of London. The external treatment only hints at the less orthodox internal arrangement. The roof garden is sited on the two shifted bedroom nests, resulting in one section of the garden being at lower elevation, creating a more intimate and private space or outdoor room. The other garden is higher, with a lower parapet allowing for an open view of the skyline of London. The plan and section are as open as possible in order to maximise the vertical connection between the ground floor, first floor, second floor and the third floor's roof garden with its greenery and views.

The digital three-dimensional model is the single source of all the project information, ranging from the early stages of design to the manufacturing and construction of the project. Initially the model was set up as a series of relationships between the different programmes, constrained by the geometry of the property, the maximum allowed pitch of the stairs, and the nominal area requirements for different activities. These relationships have been and still can be manipulated through a series of material, structural and formal strategic decisions. Data that is contained within these relationships is allowed to affect the model, allowing new patterns of organisation to arise. At this stage the digital model is entirely parametric, with all volumes and key relationships able to be altered on the fly. All two-dimensional information in terms of drawings and schedules is extracted directly from the model, as it is ready for immediate output as prototypes or as various kinds of simulations. The intention is that the complex geometry of the nests will be manufactured directly from the digital model with an extremely high precision, allowing for the minimal tolerances that are defined by the project constraints.

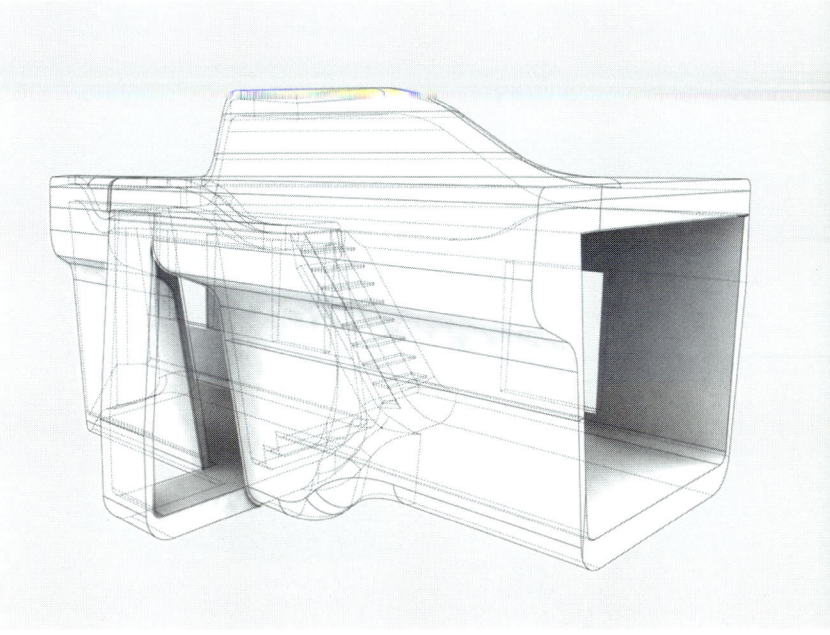

The group's aim was to develop a higher conceptual consistency in the smaller projects carried out within Urban Future Organization and its collaborations. Through Nested House V1.03, it discovered the tools and techniques for the expedient delivery of highly complex projects on a small scale. The 'single building model' paradigm technology is very much capable of delivering projects with advanced geometries and constraints, and is a valuable asset in assisting with the delivery of highly complex projects with fewer resources than conventional techniques of representation and manufacture would allow for.

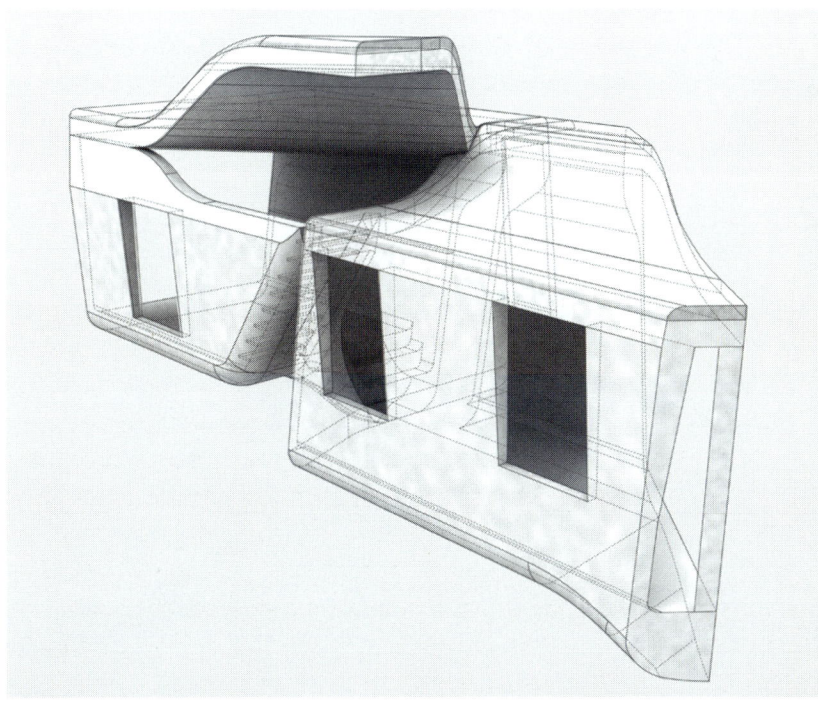

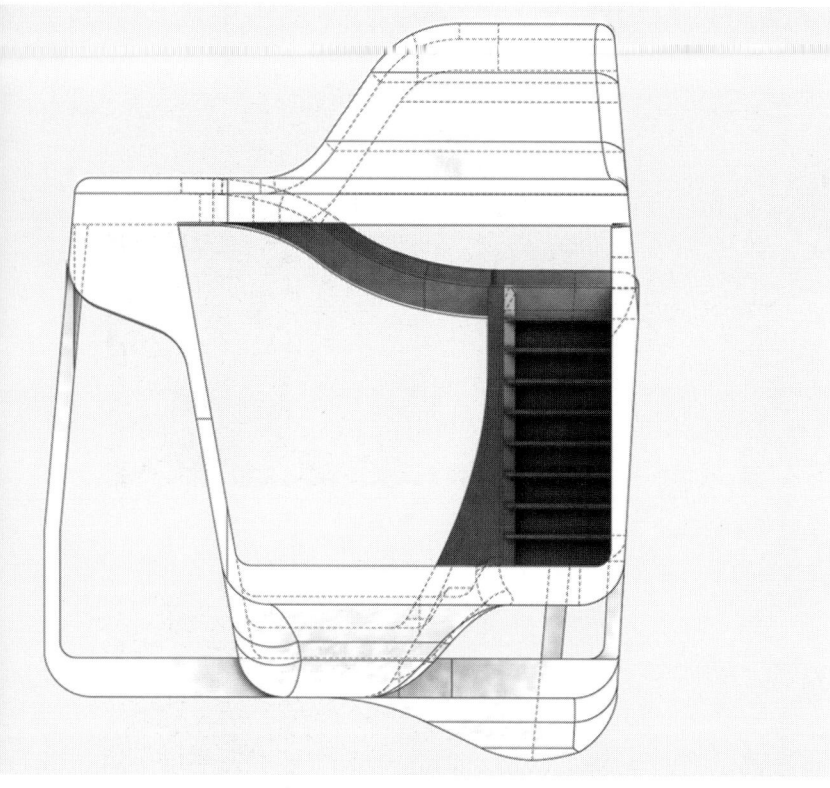

Project team
Urban Future Organization:
Artur Ferreira Viveiros
Dirk Anderson
Jonas Lundberg
Steve Hardy

Structure
Daniel Bosia

Technical Consultants
Michael Chadwick and Katja Penz

Project cost
£55,000

Start Date
Summer 2004

Urban Future Organization is a transnational networked architectural practice with a growing international reputation. A small group of its members is actively pursuing the integration of its ideas about data-driven design techniques on a live small-scale residential project in Hackney, London. In the realisation of the Nested House V1.03, it is exploring current developments in parametric design and the 'single building model' paradigm, material and structural research, as well as digital techniques of representation, production and manufacturing that have an effect on the building industry's ability to handle complex design problems.

Science Centre Wolfsburg

Zaha Hadid Architects

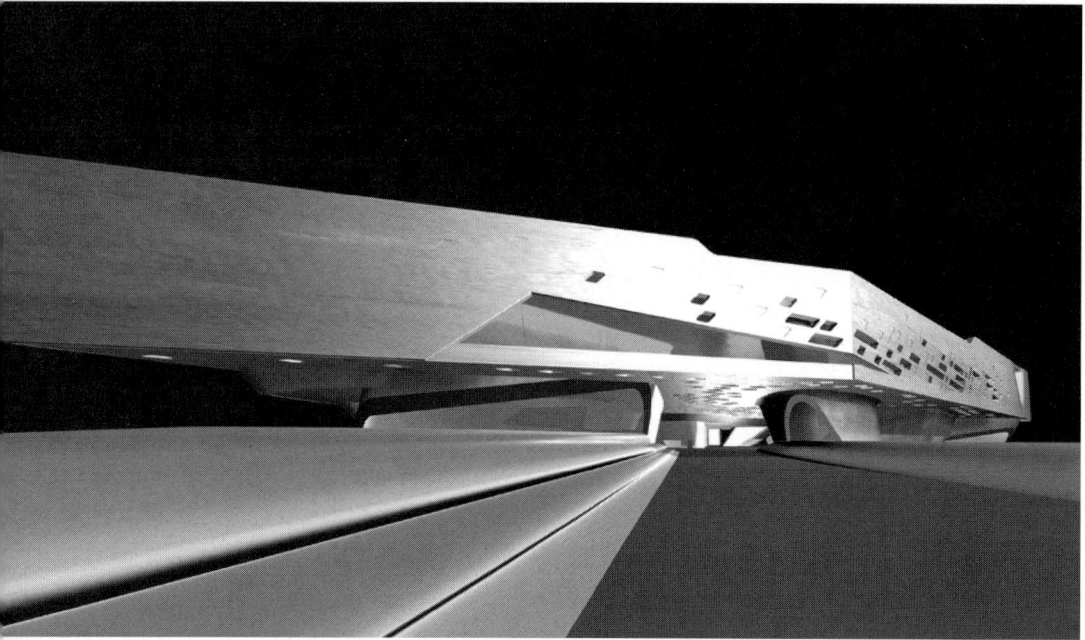

The Science Centre, the first of its kind in Germany, is conceived as a mysterious object giving rise to curiosity and discovery. The visitor is faced with a degree of the complexity and strangeness, which is ruled by a very specific system.

Located at a special site in the town – along the chain of culturally important buildings from Aalto, Scharoun and Schweger, the building is a connecting link to the new *Volkswagen-Town*, while as an effective urban mass, it closes the northern edge of the inner city along the Bahnhofsstrasse.

The big scale of the surrounding is continued while on the ground level the massiveness and enclosure of the block is dissolved and - on the base of visual axis – made porous. The area in front of the station grows wider inside the building. At this complex point of intersection the building is connected with the inner city multiple directions of movement which continue through.

The axis of important cultural buildings is drawn inside the Science Center split like a view through a kaleidoscope and finally spread in many directions towards the VW-Town.

The project is based on an unusual volumetric structural logic. The floors are neither piled above each other nor could they be seen as a hall with a large roof spanning from one side to the other.

A big volume is supported and also structured by funnel-shaped cones turned inside and out of the box above it. Through some of these funnels the interior of the box is accessible - others are used to lighten the space inside, some of them house necessary functions.

Their figure was derived from the surrounding primary urban axes. These directions then develop and organically shape the building in relation to the functions inside. Consequently, one funnel becomes the main entrance, one the lecture hall, three of them fuse to become a big exhibition space underneath the main concourse level.

An alien but simultaneously coherent crater landscape comes into existence. The public way of the bridge leads like a wormhole through the interior of the building and- like on the ground - inside and outside melt together getting through each other.

The strategy of strangeness and fusion is continued within the choice of materials. The esthetical effect of dealing with smooth, porous, acoustical damped materials and different surfaces should be the creation of a stimulating virgin territory, of a world which still has to be discovered.

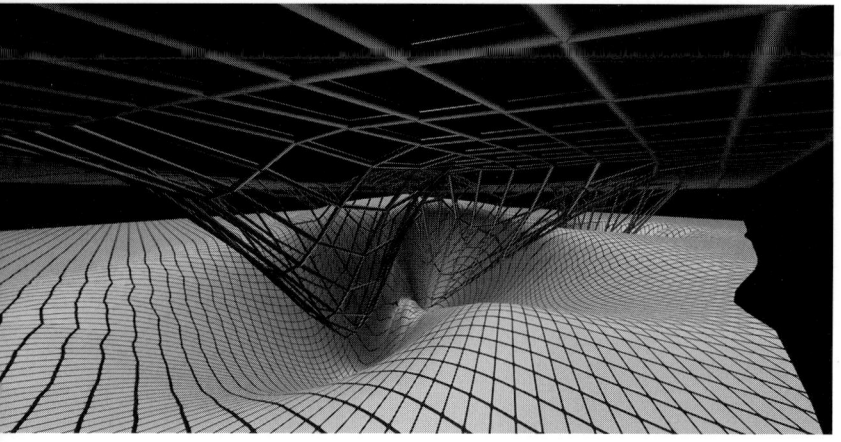

The lighting for the Wolfsburg Science Center is used as an architectural tool to allow flexibility for changing exhibitions as well as establishing a visual guiding system through the building.

To achieve a high level of flexibility the building service layers will be combined creating a grid on the main level, which becomes the service module.

This service module is the location that incorporates power outlets for lighting, HVAC, etc. As a hidden pattern it is integrated into the architectural floor plan and gains increasingly more importance as the building ages.

Future exhibition designers can easily identify the "hidden pattern" of the service grid when exhibitions need to change. The ceiling is simplified and the organized services allow simultaneously a vast open space as well as the location of temporary walls where needed.

Power will be channeled up to the ceiling to form a secondary service grid, as required. The use of darkness will be a key to the unique experience of the Science Center. Light and shadow offer the opportunity to provide a visual guiding

system through the building by creating paths of light and focal points.

The overall brightness of the interior should be minimized in order to achieve a more dynamic contrast to the highlighted exhibits, thus creating the moments of astonishment and discovery. The visitor will intuitively follow the path of illuminated focal points.

The facades can be used for projection and for watching the activity of the people within the building from outside through a kind of screen.

In addition, a smooth carpet of light will be provided underneath the building while reflected light catches the underside and sculptures the truncated volumes. Increased illumination will be used to draw visitors to the entrance areas.

Other technical lighting requirements for the Science Centre include the creation of different zones based on the speed and scale of the approach to the building (via pedestrians/cars/train).

Reinforced concrete has been chosen for the structure because it can be easily adapted to the building's free-flowing forms. The floor and roof structures are two-way spanning, waffles slabs, supported on the reinforced concrete shear walls around the workshops, auditorium, main entrance and administrative areas. In order to maximize the repetition of formwork and reinforcement, it is envisaged that the waffles will be on a regular grid. The waffles will be parallelograms, so that the concrete ribs in between will intersect at an acute angle, to follow the visual axis of the building. Due to the long spans at roof and main exhibition level, it is anticipated that the structure will be supported on piled foundations.

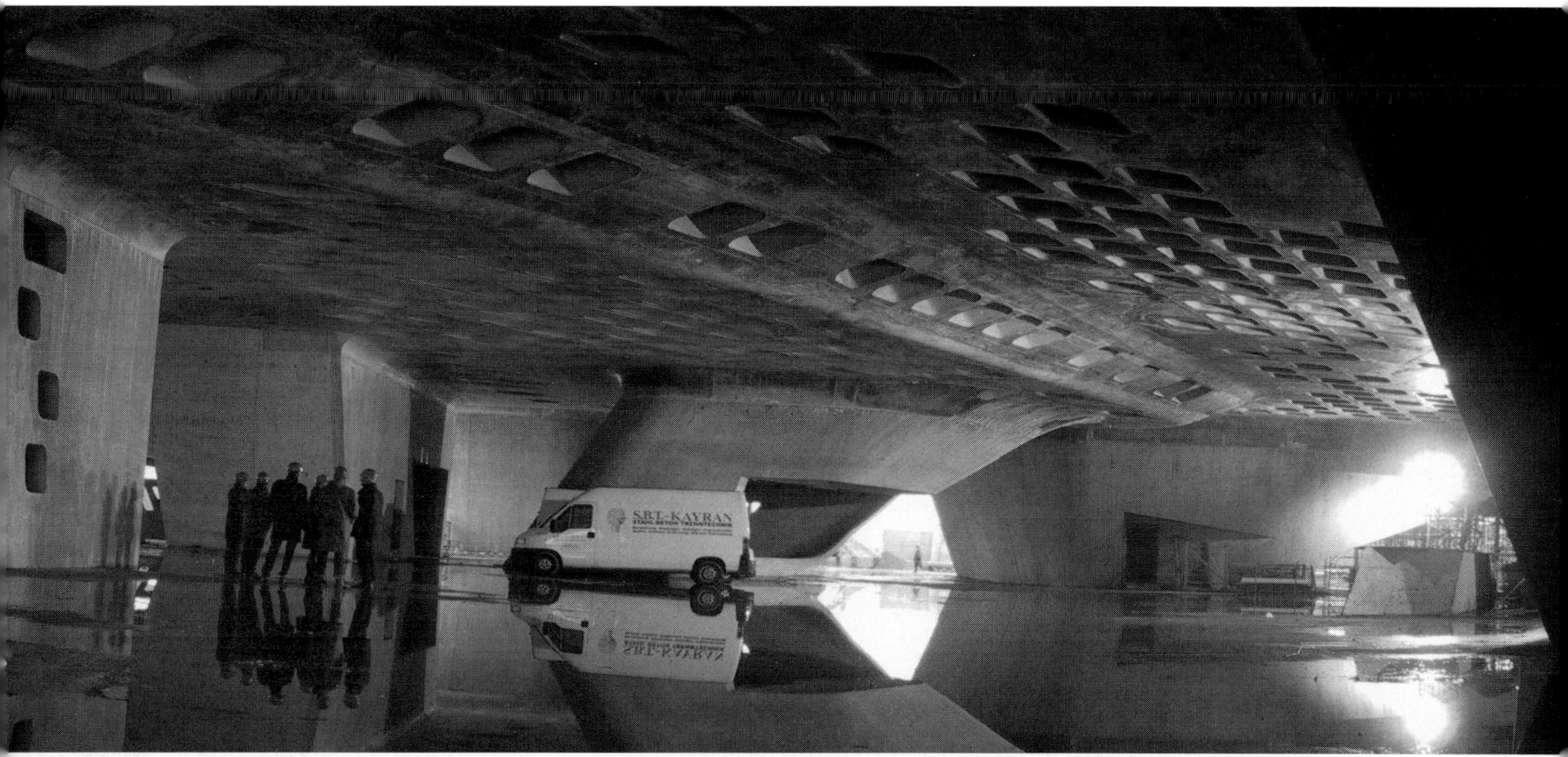
S.B.T.-KAYRAN
STAHL-BETON TRENNTECHNIK

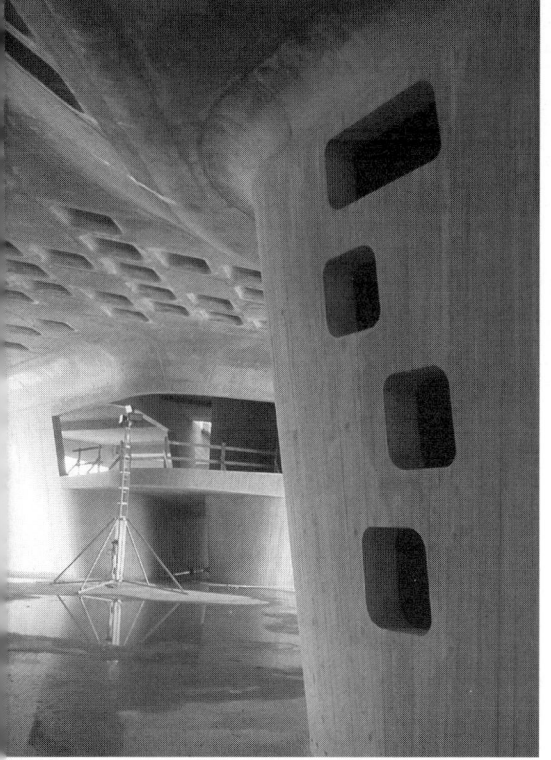

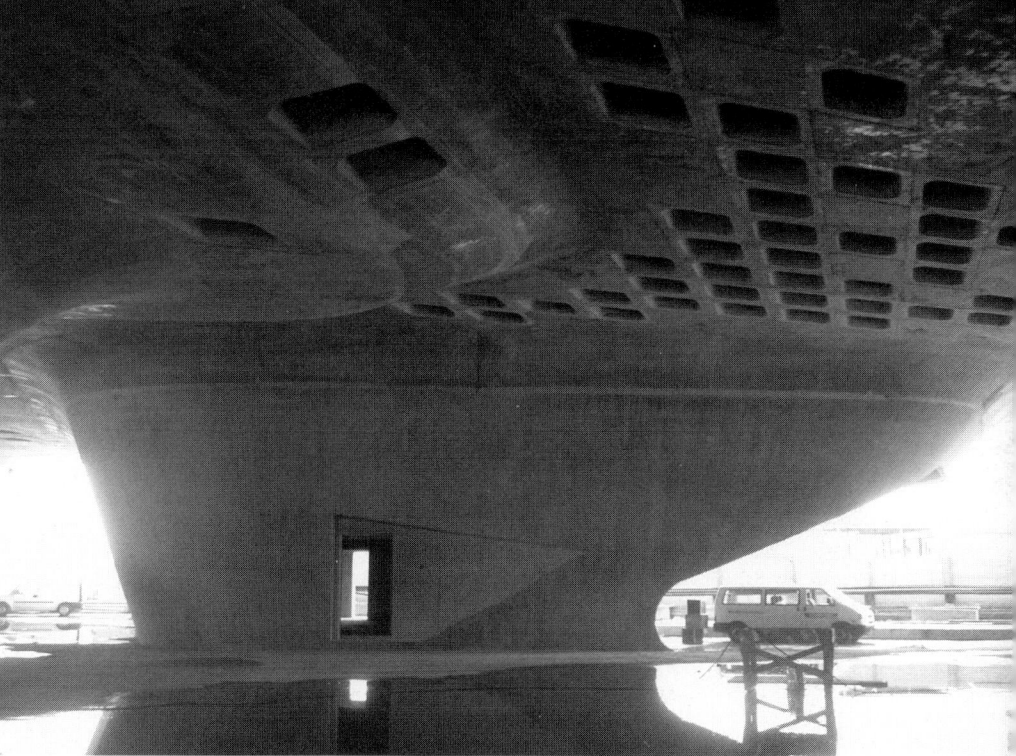

Project Credits

Richard Rogers Partnership - Antwerp Law Courts
Architect: Richard Rogers Partnership
Cladding Contractor: Permasteelisa; thv Gevelcombinatie
Antwerpen
Steelwork Contractor: Lemants
Timber Contractor: Merk
Roofing Contractor: ME Construct

Herzog & de Meuron / Arup - Beijing Olympic Stadium
A joint design consortium involving Herzog and
de Meuron Architects, ArupSport and The China
Architecture and Research Group Beijing.

Brookes Stacey Randall - Ballingdon Bridge
Architect: Brookes Stacey Randall
Engineers: Arup
Site Engineers: Suffolk Highways Engineering Consultancy
Contractor: Costain
Lighting: Equation Lighting Design
Model: Jonathan Friedman, Thomas Seebohm and Robert
Aitcheson/ Waterloo Integrated Centre for Visualization,
Design and Manufacturing

dRMM - Kingsdale School
Architects: dRMM
Structural Engineer: Michael Hadi Associates
Service Engineer: Fulcrum
Main Contractor: Galliford Try
Steelwork Subcontractor: SH Structures

Servo - Lattice Archipelogics (2002)
Servo Principals: David Erdman, Marcelyn Gow, Ulrika
Karlsson, Chris Perry
servo design team: Daniel Norell, Clare Olsen, Jonas
Runberger
Fabrication Consultants: CARAN
The Responsive Field of Lattice Archipelogics interactive
lighting system designed by:
Smart Studio / Interactive Institute - design team: Ingvar
Sjöberg, Tobi Schneidler, Fredrik Petersson, Olof Bendt,
Magnus Jonsson, Pablo Miranda

RPBW / Arup - Nasher Sculpture Gallery, Dallas
Architect: Renzo Piano Building Workshop
Engineers: Arup

Urban Future Organization - Nested House 1.03
Architects: Urban Future Organization
Structural Consultant: Daniel Bosia
Technical Consultants: Michael Chadwick, Katja Penz

Philip Beesley + Diane Willow – Orgone Reef
Computing Systems Design: Steven Wood
Design Assistants
Jonathan Tyrrell, Coryn Kempster, Farid Noufailly, Will
Elsworthy, Alex Josephson, Luanna Lam, Thomas-
Bernard Kenniff, Alex Lukachko, Vincent Hui, Kirsten
Douglas, Daniel Hall, Nancy Gibson
Prototyping
University of Waterloo's Integrated Centre for
Visualization, Design and Manufacturing
Installation: collaboration with students of University
of Waterloo School of Architecture, Erin Corcoran
coordinator

MIT / RMIT / Gehry Partners LLP - Paramorph
Reading Room.
Initial Design: Dominique Ng
Detailed Resolution Team: RMIT : Julian Canterbury,
Alison Fairley / MIT: Joy Hou, Xiaoyi Ma, Kyle Steinfeld
Work Continued at RMIT: Mark Burry, Julian
Canterbury, Alison Fairley
Managed by: MIT : Larry Sass / RMIT: Mark Burry /
Gehry Partners: Jim Glymph and Dennis Shelden

Achim Menges - Postagriculture
Architect: Achim Menges

RMIT - Sagrada Familia
Architect Coordinator and Director: Jordi Bonet, Temple
Sagrada Família, Barcelona
Project Architect: Jordi Faulí, Temple Sagrada Família,
Barcelona
Design and Documentation: SIAL, RMIT University,
Melbourne, Australia
Project Management: Ramon Espell
Stonemasons: Manuel Malló, Talleres Malló
Acknowledgements:
The Investigation of Gaudí's final design models for the
Sagrada Família church was supported by a Discovery
grant from the Australian Research Council 2000-2002.

KieranTimberlake Associates - SmartWrap
Architects: KieranTimberlake
Engineers: CVM and Buro Happold
Industrial Collaborators: DuPont, Skanska, ILC Dover and
Bosch Rexroth

Foster & Partners - Swiss Re
Architect: Foster and Partners
Structural Engineer: Arup
Main Contractor: Skanska Construction UK
Mechanical & Electrical Engineering: Hilson Moran
Partnership
Information Technology: PTS
Lighting: Speirs and Major

Enric Miralles / RMJM / Malling Precast - Scottish
Parliament Building
Architect: Enric Miralles / RMJM: Kevin Grubb
Malling Pre-cast: David Shillito
Patterns and Moulds: Gary Lucas

ONL - Web of North Holland, Floriade,
Haarlemmermeer, The Netherlands
Architects: ONL [Oosterhuis and Lénárd]
Engineers: ONL, D3BN
Construction Engineers, Meijers Staalbouw BV

D-rive - Kinetetras, Barbican Complex, London
D_rive: Anat Stern, Michael Davis, Steve Hatzellis

Piercy Conner Architects / Price & Myers
- Brooks Road
Architects: Piercy Conner Architects
Engineers: Price & Myers

Zaha Hadid Architects - Rosenthal Centre for
Contemporary Art, Cincinnati
Architects: Zaha Hadid Architects

Nio - Hoofddorp's Spaarne Hospital (Fluid Vehicle)
Project architect: Maurice Nio
Design team: Henk Bultstra, Mirjam Galjé, Hans
Larsen, Jaakko van 't Spijker
Contractor: Ooms Bouwmaatschappij bv
Structural Engineer: Ingenieursbureau Zonneveld &
Engiplast

f-u-r - ephemeralMATTER
Architects: f-u-r

Diller & Scofidio - EyeBeam Atelier/ Museum of Art
and Technology
Architects: Diller & Scofidio
Animation: Matthew Johnson with D-Box
Associate Architects: Helfand Myerberg
Guggenheimer

magma - Munster Civic Centre
Architects: magma architecture
With Ross Anderson, Stephane Canuccini, Francisco
Rosé Diaz, Yohko Mizushima, Nikola Schulz
Model: Bertolt Schulz

Future Systems / Arup - Selfridges Footbridge
Architects: Future Systems
Engineers: Arup
Research Team: Arup and RMIT SIAL

Zaha Hadid Architects - Wolfsburg Science Centre
Architects: Zaha Hadid Architects + Mayer Bahrle
Freie Architekten BDA, Germany
Structural Engineers: Adams Kara Taylor, Tokarz
Freirichs Leipold
Services Engineers: NEK, Buro Happold

Ocean D - Palisades Glacier Mountain Hut
Design and Concept: Ocean D

**Veech Media - Tissue Engineering
for Museum Quarters**
Design: Veech Media
Technical Consultant: Professor Tom Barker - b
consultants, London

Bruce Gernand - Star and Cloud
Sculptor: Bruce Gernand

**PTW / Arup - Water Cube National Swimming
Centre, Beijing**
A joint design consortium involving Pendle
Thorp Walker, Arup and The China Construction
Engineering Corporation

Alsop Architects / Architen Landrell - Spiky Pod,
Queen Mary Westfield College
Architect: Alsop Architects/AMEC Capital Projects
Engineer: Adams Kara Taylor
Fabricator: Architen Landrell
Fabric Patterning: Tensys

Digital Fabricators
Michael Stacey, curator,
with London Metropolitan University
Digital Fabricators Research Group (DFRG)
Michael Scoones, Building Centre Trust
Jackson Hunt, Building Centre Trust

AIA/ACADIA Fabrication Conference

Chairs
Philip Beesley, University of Waterloo
Nancy Cheng, University of Oregon
Shane Williamson, University of Toronto

Megan Torza, Conference Coordinator
Farid Noufailly, Administrative Coordinator
Alex Josephson, Assistant Coordinator
Nadine Beaulieu, Assistant Coordinator
Coryn Kempster, Assistant Coordinator

Exhibition team
Mary Misner, Cambridge Galleries Director
Vincent Hui, Project Manager
John McMinn, Architecture Cambridge Coordinator

Catalogue Design and Production
Philip Beesley
Farid Noufailly
Jonathan Tyrrell
Shane Williamson
Marianne Magus